.2.73

STATISTICS:

PROBLEMS AND SOLUTIONS

A Complete Course in Statistics

by

J. Murdoch BSc, ARTC, AMIProdE

and

J. A. Barnes BSc, ARCS

STATISTICS, PROBLEMS AND SOLUTIONS
BASIC STATISTICS, LABORATORY INSTRUCTION MANUAL
STATISTICAL TABLES FOR SCIENCE ENGINEERING
BUSINESS STUDIES AND MANAGEMENT

STATISTICS:
PROBLEMS AND SOLUTIONS

J. Murdoch, BSc, ARTC, AMIProdE

Head of Statistics and Operational Research Section,
Cranfield Institute of Technology

and

J. A. Barnes, BSc, ARCS

Lecturer in Statistics and Operational Research,
Cranfield Institute of Technology

MACMILLAN

© J. Murdoch and J. A. Barnes 1973

All rights reserved. No part of this publication may be reproduced or transmitted, in any form or by any means, without permission

First published 1973

Published by
THE MACMILLAN PRESS LTD
London and Basingstoke
Associated companies in New York Toronto
Melbourne Dublin Johannesburg and Madras

SBN 333 12017 5

Printed in Great Britain by
The Whitefriars Press Ltd., London and Tonbridge

Preface

Statistics is often regarded as a boring, and therefore difficult, subject particularly by those whose previous experience has not produced any real need to understand variation and to make appropriate allowances for it. The subject can certainly be presented in a boring way and in much advanced work can be conceptually and mathematically very difficult indeed.

However for most people a simple but informed approach to the collection, analysis and interpretation of numerical information is of tremendous benefit to them in reducing some of the uncertainties involved in decision making. It is a pity that many formal courses of statistics appear to frighten people away from achieving this basic attitude usually through failing to relate the theory to practical applications.

This book, whose chapters each contain a brief summary of the main concepts and methods, is intended to show, through worked examples, some of the practical applications of simple statistical methods and so to stimulate interest. In order to establish firmly the basic concepts, a more detailed treatment of the theory is given in chapters 1 and 2. Some examples of a more academic nature are also given to illustrate the way of thinking about problems. Each chapter contains problems for the reader to attempt, the solutions to these being discussed in some detail, particularly in relation to the inferences that can validly be drawn even in those cases where the numbers have been put into the correct 'textbook formula' for the situation.

This book will not only greatly assist students to gain a better appreciation of the basic concepts and use of the theory, but will also be of interest to personnel in industry and commerce, enabling them to see the range of application of basic statistical concepts.

For the application of basic statistics, it is essential that statistical tables are used to reduce the computation to a minimum. The tables used are those by the authors, *Statistical Tables,* a companion volume in this series of publications on statistics. The third book, *Basic Statistics: Laboratory Instruction Manual,* designed to be used with the Cranfield Statistical Teaching Aids is referred to here and, in addition, some experiments are suggested for the reader to perform to help him understand the concepts involved. In the chapters of this book references to *Statistical Tables for Science, Engineering and Management* are followed by an asterisk to distinguish them from references to tables in this book.

The problems and examples given represent work by the authors over many years and every attempt has been made to select a representative range to illustrate the basic concepts and application of the techniques. The authors would like to apologise if inadvertently examples which they have used have previously been published. It is extremely difficult in collating a problem book such as this to avoid some cases of duplication.

It is hoped that this new book, together with its two companion books, will form the basis of an effective approach to the teaching of statistics, and certainly the results from its trials at Cranfield have proved very stimulating.

J. Murdoch
Cranfield J. A. Barnes

Contents

List of symbols

a	constant term in linear regression
b	regression coefficient
c	scaling factor used in calculating mean and variance
E_i	expected frequency for χ^2 goodness-of-fit test
E_{ij}	expected frequency in cell ij of contingency table
f_i	frequency of ith class
F	variance ratio
H_0	null hypothesis
H_1	alternative hypothesis
m	mean of poisson distribution
μ_r^1	rth moment of a distribution about the origin
μ_r	rth moment of a distribution about its mean
n	number of observations and/or number of trials in probability theory
O_i	observed frequency for χ^2 goodness-of-fit test
O_{ij}	observed frequency in ijth cell of contingency table
p	probability
$P(A)$	probability of an event A
$P\int_x^n$	number of permutations of n objects taken x at a time
$P(A/B)$	conditional probability of A on assumption that B has occurred
$E[x]$	expected value of variate, x
r	sample correlation coefficient
s'	standard deviation of a sample
s^2	unbiased sample estimator of population variance
t	Student's 't'
u	coded variable used in calculating mean and variance of sample
also†	
u	standardised normal variate
x_i	value of variate
y_i	value of dependent variable corresponding to x_i in regression
Y_i	estimated value of dependent variable using the regression line

† Little confusion should arise here on the use of the same symbol in two different ways. Their use in both these areas is too standardised for the authors to suggest a change.

Greek Symbols

μ	population mean
σ^2	population variance
χ^2	sum of the squares of standardised normal deviates
ν	number of degrees of freedom
α	magnitude of risk of 1st kind or significance level
β	magnitude of risk of 2nd kind or $(1-\beta)$ is the power of the test
π	proportion of a population having a given attribute
ϵ	standard error

Note: α and β are also used as parameters of the population regression line $\eta = \alpha + \beta\,(x_t - x)$ but again no confusion should arise.

Mathematical Symbols

$\sum_{i=1}^{n}$	summation from $i = 1$ to n
e	exponential 'e', the base of natural logarithms
$\doteqdot$	approximately equal to
$b \simeq \beta$	the sample statistic b is an estimate of population parameter β
$x > y$	x is greater than y
$x \geqslant y$	x is greater than or equal to y
$x < y$	x is less than y
$x \leqslant y$	x is less than or equal to y
$\mathcal{C}_x^n$ or $\binom{n}{x}$	number of different combinations of size x from group of size n
$x!$	factorial $x = x(x-1)(x-2)\ldots 3 \times 2 \times 1$
P_x^n	number of permutations of n objects taken x at a time

Note: The authors use $\binom{n}{x}$ but in order to avoid any confusion both are given in the definitions.

1 Probability theory

1.1 Syllabus Covered

Definition and measurement of probability; addition and multiplication laws; conditional probability; permutations and combinations; mathematical expectation; geometric probability; introduction to hypergeometric and binomial laws.

1.2 Résumé of Theory and Basic Concepts

1.2.1 *Introduction*

Probability or chance is a concept which enters all activities. We speak of the chance of it raining today, the chance of winning the football pools, the chance of getting on the bus in the mornings when the queues are of varying size, the chance of a stock item going out of stock, etc. However, in most of these uses of probability, it is very seldom that we attempt to measure or quantify the statements. Most of our ideas about probability are intuitive and in fact probability is a quantity rather like length or time and therefore not amenable to simple definition. However, probability (like length or time) can be measured and various laws set up to govern its use.

The following sections outline the measurement of probability and the rules used for combining probabilities.

1.2.2 *Measurement of Probability*

Probability is measured on a scale ranging from 0 to 1 and can take any value inside this range. This is illustrated in figure 1.1.

The probability p that an event (A) will occur is written

$$P(A) = p \quad \text{where} \quad 0 \leqslant p \leqslant 1$$

1	Probability that you will die one day (absolute certainty)
$\frac{1}{2}$ or 0.5	Probability that an unbiased coin shows 'heads' after one toss
$\frac{1}{6}$ or 0.167	Probability that a die shows 'six' on one roll
0	Probability that you will live forever (absolute impossibility)

Figure 1.1. Probability scale.

It will be seen that on this continuous scale, only the two end points are concerned with deductive logic (although even here, there are certain logical difficulties with the particular example quoted).

On this scale absolute certainty is represented by $p = 1$ and an impossible event has probability of zero. However, it is between these two extremes that the majority of practical problems lie. For instance, what is the chance that a machine will produce defective items? What is the probability that a machine will find the overhead service crane available when required? What is the probability of running out of stock of any item? Or again, in insurance, what is the chance that a person of a given age will survive for a further year?

1.2.3 *Experimental Measurement of Probability*

In practice there are many problems where the only method of estimating the probability is the following

$$\text{probability of event } A, P(A) = \frac{\text{total occurrences of the event } A}{\text{total number of trials}}$$

For example, what is the probability of an item's going out of stock in a given period?

Measurement showed that 189 items ran out in the period out of a total number of stock items of 2000, therefore the estimate of probability of a stock running out is

$$P(A) = \tfrac{189}{2000} = 0.0945$$

Again, if out of a random sample of 1000 men, 85 were found to be over 1.80 m tall, then

$$\text{estimate of probability of a man being over 1.80 m tall} = \tfrac{85}{1000} = 0.085.$$

1.2.4 *Basic Laws of Probability*

1. *Addition Law of Probability*

This law states that if A and B are mutually exclusive events, then the probability that either A or B occurs in a given trial is equal to the sum of the separate probabilities of A and B occurring.

In symbolic terms this law can be shown as

$$P(A \text{ or } B) = P(A) + P(B)$$

This law can be extended by repeated application to cover the case of more than two mutually exclusive events.

Thus $P(A \text{ or } B \text{ or } C \text{ or } \ldots) = P(A) + P(B) + P(C) + \ldots$

The events of this law are mutually exclusive events, which simply means that the occurrence of one of the events excludes the possibility of the occurrence of any of the others on the same trial.

For example, if in a football match, the probability that a team will score 0 goals is 0.50, 1 goal is 0.30, 2 goals is 0.15 and 3 or more goals is 0.05, then the probability of the team scoring either 0 or 1 goals in the match is

$$P(0 \text{ or } 1) = P(0) + P(1) = 0.50 + 0.30 = 0.80$$

Also, the probability that the team will score at least one goal is

$$P(\text{at least one goal}) = P(1) + P(2) + P(3 \text{ or more}) = 0.30 + 0.15 + 0.05 = 0.50$$

Any event either occurs or does not occur on a given occasion. From the definition of probability and the addition law, the probabilities of these two alternatives must sum to unity. Thus the probability that an event does *not* occur is equal to

$$1 - (\text{probability that the event does occur})$$

In many examples, this relationship is very useful since it is often easier to find the probability of the complementary event first.

For example, the probability of a team's scoring at least one goal in a football match can be obtained as

$$P(\text{at least 1 goal}) = 1 - P(0 \text{ goals}) = 1 - 0.50 = 0.50$$

as before.

As a further example, suppose that the probabilities of a man dying from heart disease, cancer or tuberculosis are 0.51, 0.16 and 0.20 respectively. The probability that a man will die from heart disease or cancer is $0.51 + 0.16 = 0.67$. The probability that he will die from some cause other than the three mentioned is $1-(0.51 + 0.16 + 0.20) = 0.13$; i.e., 13% of men can be expected to die from some other cause.

However, consider the following example. Suppose that of all new cars sold, 40% are blue and 30% have two doors. Then it cannot be said that the probability of a person's owning either a blue car or a two-door car is 0.70 (= 0.40 + 0.30) since the events (blue cars and two-door cars) are not mutually exclusive, i.e., a car can be both blue and have two doors. To deal with this case, a more general version of the addition law is necessary. This may be stated as

$$P(A \text{ or } B \text{ or both}) = P(A) + P(B) - P(A \text{ and } B)$$

The additional term, the probability of events A and B both occurring together on a trial is obtained using the multiplication law of probabilities.

2. *Multiplication Law of Probability*

This law states that the probability of the combined occurrence of two events A and B is the product of the probability of A and the conditional probability of B on the assumption that A has occurred.

Thus $P(A \text{ and } B) = P(AB) = P(A) \times P(B/A)$

where $P(B/A)$ is the conditional probability of event B on the assumption that A occurs at the same time (see list of symbols, page xi).

$$P(AB) \text{ is also given by } P(B) \times P(A/B)$$

While this law is usually defined as above for two events, it can be extended to any number of events.

3. *Independent Events*

Events are defined as independent if the probability of the occurrence of either is not affected by the occurrence or not of the other. Thus if A and B are independent events, then the law states that the probability of the combined occurrence of the events A and B is the product of their individual probabilities. That is

$$P(AB) = P(A) \times P(B)$$

Many people meeting the ideas of probability for the first time find difficulty in deciding whether to add or multiply probabilities. If the problem (or part of it) is concerned with *either* event A *or* event B occurring, then *add* probabilities; if A *and* B must both occur (at the same time or one after the other), then *multiply* probabilities. Consider the following example with the throwing of two dice illustrating the use of the two basic laws of probability.

Examples

1. In the throw of two dice, what is the probability of obtaining two sixes?

One of the dice must show a six and the other must also show a six. Thus the required probability (independent events) is

$$p = \tfrac{1}{6} \times \tfrac{1}{6} = \tfrac{1}{36}$$

2. In the throw of two dice, what is the probability of a score of 9 points?

Here we must consider the number of mutually exclusive ways in which the score 9 can occur. These ways are listed below

Dice A	3		4		5		6
and	|	or	|	or	|	or	|
Dice B	6		5		4		3

The probability of any of these four possible arrangements occurring is equal to, as before $(\frac{1}{6} \times \frac{1}{6}) = \frac{1}{36}$

Thus the probability that two dice show a total score of 9 is equal to

$$\frac{1}{36} + \frac{1}{36} + \frac{1}{36} + \frac{1}{36} = \frac{1}{9}$$

3. In marketing a product, records show that on average one call in 10 results in making a sale to a potential customer. What is the probability that a salesman will make two sales from any two given calls?

Assuming the events (sales to different customers) to be independent, use of the multiplication law gives the probability of making two sales in two calls as

$$0.1 \times 0.1 = 0.01.$$

As an extension of this example, what is the probability of making at least one sale in five calls? The easiest way to calculate this probability is to note that the event complementary to making one or more sales is not to make any sales. Using the multiplication law gives the probability of making no sales in five calls as

$$0.9 \times 0.9 \times 0.9 \times 0.9 \times 0.9 = 0.9^5 = 0.5905$$

$$P(\text{at least one sale in five calls}) = 1 - 0.5905 = 0.4095$$

These basic laws for combining probability may be used to answer such questions as how many calls must be planned so that there is a high probability, say 95% or 99%, of making at least one sale or of making at least two sales, etc. Or, again, what is the probability that it will need more than, say eight calls to make two sales?

As an example, suppose that the probability of making at least one sale in n calls is to be at least 0.95. What is the smallest value of n which will achieve this?

Turning the problem round gives that the probability of making no sales in n calls is to be at most 0.05 and thus

$$0.9^n \leqslant 0.05$$

The smallest value of n which satisfies this requirement is 29. This means that if the salesman schedules 29 customer calls every day, he will make *at least* one sale on just over 95% of days in the long run. Conversely on just under one day in 20 he will receive no orders as a result of any of his 29 visits. The average daily number of sales made will be 2.9.

The example of the addition law where the events (a car being blue and a car having two doors) were not mutually exclusive can now be completed.

The probability that a randomly chosen car is either blue or has two doors or is a two-door blue car is given by

$$P(\text{blue}) + P(2 \text{ doors}) - P(\text{blue and 2 doors})$$

$$= 0.4 + 0.3 - (0.4 \times 0.3)$$

$$= 0.70 - 0.12 = 0.58$$

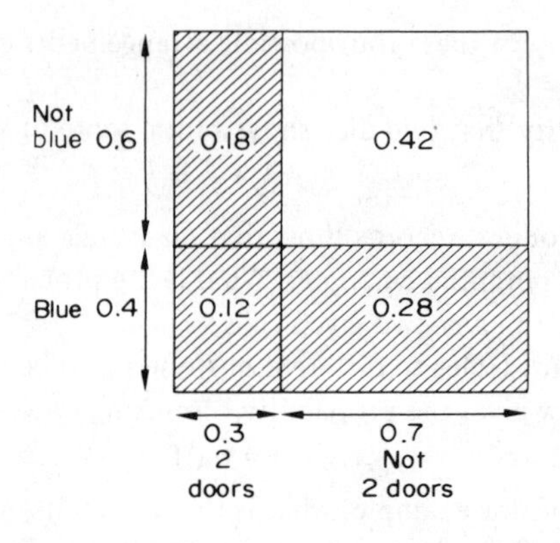

Figure 1.2

This result is valid on the assumption that the number of doors that a car has is *not* dependent on its colour.

Figure 1.2 illustrates the situation. The areas of the four rectangles within the square represent the proportion of all cars having the given combination of colour and number of doors. The total area of the three shaded rectangles is equal to 0.58, the proportion of cars that are either blue *or* have two doors *or* are two-door blue cars.

1.2.5 *Conditional Probability*

In many problems and situations, however, the events are neither independent nor mutually exclusive, and the general theory for conditional probability will be outlined here.

Before considering conditional probability in algebraic terms, some simple numerical examples will be given.

If one card is drawn at random from a full pack of 52 playing cards, the probability that it is red is 26/52. Random selection of a card means that each of the 52 cards is as likely as any of the others to be the sampled one.

If a second card is selected at random from the pack (without replacing the first), the probability that it is red depends on the colour of the first card drawn. There are only 51 cards that could be selected as the second card, all of them having an equal chance. If the first had been black, there are 26 red cards available and the probability that the second card is red is therefore 26/51 (i.e., conditional upon the first card being black).

Similarly, if the first card is red, the probability that the second is also red is 25/51.

The process can be continued; the probability that a third card is red being 26/50, 25/50 or 24/50 depending on whether the previous two cards drawn were both black, one of each colour (drawn in either order) or both red.

Most practical sampling problems are of this 'sampling without replacement' type and conditional probabilities need to be taken into account. There are, however, suitable approximations which can often be used in practice instead of working with the exact conditional values. (These methods are referred to in chapter 5.)

Using the multiplication law and the conditional probabilities discussed above, the probability that two cards, taken randomly from a full pack (52 cards), will both be red is given by

P(first card red) x P(second card red given that first was red) = $\frac{26}{52} \times \frac{25}{51} = \frac{25}{102}$

This result applies whether the two cards are taken one after the other or both at the same time.

As another example, suppose that two of the bulbs in a set of 12 coloured lights are burnt out. What is the probability of finding

(*a*) both burnt-out bulbs in the first two tested?
(*b*) one of the burnt-out bulbs in the first two tested?
(*c*) at least one of the burnt-out bulbs in the first two tested?

The solutions are

(*a*) P(first bulb tested is burnt out) = $\frac{2}{12}$
P(second bulb tested is also burnt out) = $\frac{1}{11}$
P(finding both burnt-out bulbs in first two tests) = $\frac{2}{12} \times \frac{1}{11} = \frac{1}{66}$

(*b*) P(first bulb tested is burnt out) = $\frac{2}{12}$
and P(second bulb tested is *not* burnt out) = $\frac{10}{11}$
The product of these probabilities = $\frac{2}{12} \times \frac{10}{11} = \frac{10}{66} = \frac{5}{33}$

The same result can be obtained if a burnt-out bulb is found on the second test, the first bulb being good. The two situations are mutually exclusive.
Then P(first bulb good) = $\frac{10}{12}$
P(second bulb burnt out) = $\frac{2}{11}$
Thus P(first bulb good and second burnt out) = $\frac{10}{12} \times \frac{2}{11} = \frac{10}{66}$

Either of these two situations satisfies (*b*) and the probability of at least one of the burnt-out bulbs being in the first two tested is given by their sum

$$\frac{10}{66} + \frac{10}{66} = \frac{20}{66} = \frac{10}{33}$$

(*c*) The probability of at least one burnt-out bulb being found in two tests is equal to the sum of the answers to parts (*a*) and (*b*), namely

$$\frac{20}{66} + \frac{1}{66} = \frac{21}{66} = \frac{7}{22}$$

As a check on this result, the only other possibility is that neither of the faulty bulbs will be picked out for the first two tests. The probability of this, using the multiplication law with the appropriate conditional probability, is

$$\tfrac{10}{12} \times \tfrac{9}{11} = \tfrac{15}{22}$$

The situation in part (c) therefore has probability of $1 - \tfrac{15}{22} = \tfrac{7}{22}$ as given by direct calculation.

Consider a box containing r red balls and w white balls. A random sample of two balls is drawn. What is the probability of the sample containing two red balls?

r = red balls
w = white balls

If the first ball is red (event A), probability of this event occurring

$$P(A) = \frac{r}{r+w}$$

The probability of the second ball being red (event B) given the first was red is thus

$$P(B/A) = \frac{r-1}{r+w-1}$$

since there are now only $(r-1)$ red balls in the box containing $(r+w-1)$ balls.
∴ Probability of the sample containing two red balls

$$= \frac{r}{(r+w)} \times \frac{(r-1)}{(r+w-1)}$$

In similar manner, probability of the sample containing two white balls

$$= \frac{w}{(r+w)} \times \frac{(w-1)}{(r+w-1)}$$

Also consider the probability of the samples containing one red and one white ball. This event can happen in two mutually exclusive ways, first ball red, second ball white or first ball white, second ball red.

Thus, the probability of the sample containing one white and one red ball is

$$\frac{r}{(r+w)} \times \frac{w}{(r+w-1)} + \frac{w}{(r+w)} \times \frac{r}{(r+w-1)} = \frac{2wr}{(r+w)(r+w-1)}$$

Note: Readers might like to verify that the sum of these three probabilities = 1.

Examples

1. In a group of ten people where six are male and four are female, what is the chance that a committee of four, formed from the group with random selection, comprises (*a*) four females, or (*b*) three females and one male?

(*a*) Probability of committee with four females

$$\tfrac{4}{10} \times \tfrac{3}{9} \times \tfrac{2}{8} \times \tfrac{1}{7} = 0.0048$$

(*b*) Committee comprising three females and one male. This committee can be formed in the following mutually exclusive ways

1st member	M		F		F		F
2nd member	F	or	M	or	F	or	F
3rd member	F		F		M		F
4th member	F		F		F		M

The probability of the first arrangement is

$$\tfrac{6}{10} \times \tfrac{4}{9} \times \tfrac{3}{8} \times \tfrac{2}{7} = 0.0286$$

The probability for the second arrangement is

$$\tfrac{4}{10} \times \tfrac{6}{9} \times \tfrac{3}{8} \times \tfrac{2}{7} = 0.0286$$

and similarly for the third and fourth columns, the position of the numbers in the numerator being different in each of the four cases. The required probability is thus 4 x 0.0286 = 0.114.

2. From a consignment containing 100 items of which 10% are defective, a random sample of 10 is drawn. What is the probability of (*a*) the sample containing no defective items, or (*b*) the sample containing exactly one defective item?

(*a*) Probability of no defective items in the sample only arises in one way.

∴ Probability of no defective items

$$P(0) = \tfrac{90}{100} \times \tfrac{89}{99} \times \tfrac{88}{98} \times \ldots \times \tfrac{81}{91} = 0.33$$

(*b*) Exactly one defective item in the sample can arise in 10 mutually exclusive ways as shown below

1	2	3	4	5	6	7	8	9	10	
D	G	G	G	G	G	G	G	G	G	D = defective item
G	D	G	G	G	G	G	G	G	G	G = good item
G	G	G	G	G	G	G	G	G	D	

Thus the probability of one defective in 10 items sampled is

$$P(1) = 10 \times \tfrac{10}{100} \times \tfrac{90}{99} \times \tfrac{89}{98} \times \ldots \times \tfrac{82}{91} = 0.41$$

1.2.6 *Theory of Groups*

There are two group theories which can assist in the solution and/or computation involved in probability theory (rather than the long methods used in examples in section 1.2.5).

Permutations

Groups form different permutations if they differ in any or all of the following aspects.

(1) Total number of items in the group.
(2) Number of items of any one type in the group.
(3) Sequence.

Thus

ABB, BAB are different permutations (3); *AA, BAA* are different permutations (1) and (2); *CAB, CAAB* are different permutations (1) and (2); *BAABA, BABBA* are different permutations because of (2).

Thus distinct arrangements differing in (1) and/or (2) and/or (3) form different permutations.

Group Theory No. 1

If there are n objects, each distinct, then the number of permutations of objects taken x at a time is

$$P_x^n = \frac{n!}{(n-x)!}$$

An example is the number of ways of arranging two different letters out of the word *girl.*

Here $n = 4, x = 2$

$$P_2^4 = \frac{4!}{2!} = 12$$

The arrangements are

gi, gr, gl, ir, il, rl, lr, li, ri, lg, rg, ig

Combinations

Groups form different combinations when they differ in

(1) Total number of objects in the group.
(2) Number of objects of any one type in the group.
(*Note:* Sequence does not matter.)

Thus *ABB*, *BAB*, *BBA*, are not different combinations.

Group Theory No. 2

If there are n objects, each distinct, then the number of different combinations of size x is given by

$$C_x^n \quad \text{or} \quad \binom{n}{x} = \frac{n!}{x!(n-x)!}$$

As an example, a committee of three is to be formed from five department heads. How many different committees can be formed?

$$\binom{5}{3} = \frac{5!}{3!2!} = \frac{5 \times 4}{2 \times 1} = 10$$

1.2.7 *Mathematical Expectation*

In statistics, the term *expected value* refers to the average value that a variable takes. It is often used in the context of gambling but its use is appropriate whenever we are concerned with average values.

The *expected value* is thus the mean of a distribution (see chapter 2), i.e., the average sample value which will be obtained when the sample size tends to infinity.

Suppose player A receives an amount of money M_1 if event E_1 happens, an amount M_2 if E_2 happens, . . . and amount M_n if E_n happens, where $E_1, E_2, \ldots E_n$ are mutually exclusive and exhaustive events; $P_1, P_2, \ldots P_n$ are the respective probabilities of these events. Then A's mathematical expectation of gain is defined as

$$E(M) = M_1P_1 + M_2P_2 + \ldots + M_nP_n$$

In gambling, for the game to be fair the expectation should equal the charge for playing the game. This concept is also used in insurance plans, etc. Use of this concept of expected value is illustrated in the following example.

Example

The probability that a man aged 55 will live for another year is 0.99. How large a premium should he pay for £2000 life insurance policy for one year? (Ignore insurance company charges for administration, profit, etc.)

Let s = premium to be paid

Expected return = 0 x 0.99 + £2000 x 0.01 = £20

∴ Premium s = £20 (should equal expected return)

1.2.8 *Geometric Probability*

Many problems lend themselves to solutions only by using the concept of geometric probability and this will be illustrated in this section.

Example–A Fairground Problem

At a fair, customers roll a coin onto a board made up to the pattern shown in figure 1.3. If the coin finishes in a square (not touching any lines), the number of coins the customer will win is shown in that square, but otherwise the coin is lost. If at least half of the coin is outside the board, it is returned to the player.

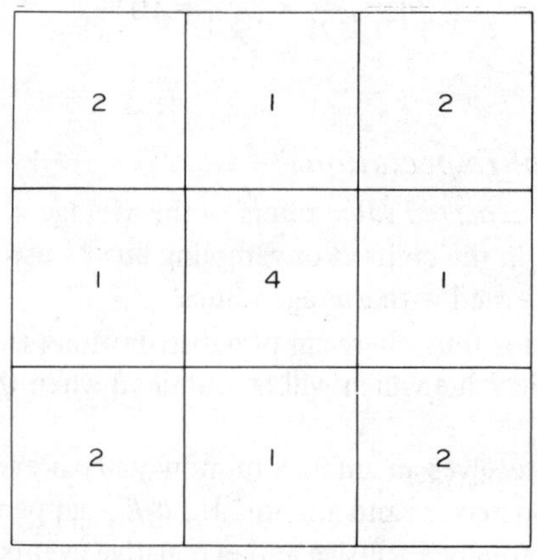

Figure 1.3

Given that the lines are 1 mm thick, the sides of the squares are 60 mm and the diameter of the coin is 20 mm what is

(*a*) the chance of getting the coin in the *4* square?
(*b*) the chance of getting the coin in a *2* square?
(*c*) the expected return per trial, if returns are made in accordance with the numbers in the squares?

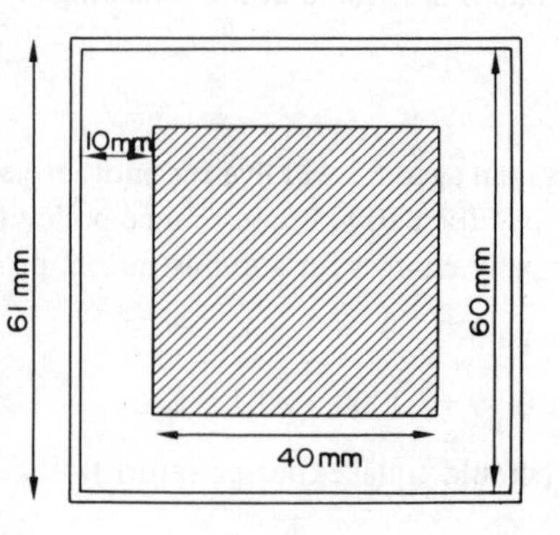

Figure 1.4

Considering one square (figure 1.4), total possible area (ignoring small edge-effects of line thickness) = $61^2 = 3721$

$\therefore$ For one square, the probability that the coin does not touch a line is

$$\frac{1600}{3721} = 0.43$$

Thus if the coin falls at random on the board

(*a*) the chance that it falls completely within the *4* square = $\frac{1}{9} \times 0.43 = 0.048$
(*b*) the chance that it falls completely within a *2* square = $\frac{4}{9} \times 0.43 = 0.191$
(*c*) the expected payout per trial is $(4 \times 0.048) + (2 \times 0.191) + (1 \times 0.191)$
$= 0.76$

Since it costs *one* coin to play, the player will lose 0.24 of a coin *per turn* in the *long run.*

1.2.9 *Introduction to the Hypergeometric Law*

The hypergeometric law gives an efficient way of solving problems where the probabilities involved are conditional.

In general form, it can be defined as follows.

Definition of Hypergeometric Law

If a group contains N items of which M are of one type and the remainder $N-M$, are of another type, then the probability of getting exactly x of the first type in a random sample of size n is

$$P(x) = \frac{\binom{M}{x}\binom{N-M}{n-x}}{\binom{N}{n}}$$

To illustrate the use of the hypergeometric law, consider example (2), page 9 again.

Here $N = 100$
$M = 10$ or number of defective items in the batch
$N - M = 90$ or number of good items in the batch

Sample size $n = 10$

$\therefore$ For (*a*) $x = 0$,

$$P(0) = \frac{\binom{10}{0}\binom{90}{10}}{\binom{100}{10}} = \frac{\dfrac{10!}{0!10!} \times \dfrac{90!}{10!80!}}{\dfrac{100!}{10! \times 90!}}$$

$$= \frac{90}{100} \times \frac{89}{99} \times \ldots \times \frac{81}{91} = 0.33$$

for (*b*) $x = 1$, $$P(1) = \frac{\binom{10}{1}\binom{90}{9}}{\binom{100}{10}} = 0.41$$

Both results are the same as before but are obtained more easily.

1.2.10 *Introduction to the Binomial Law*

Although this law will be dealt with more fully in chapter 3, it is useful to introduce it here in the chapter on probability since knowledge of the law helps in the understanding of probability.

Definition of Binomial Law

If the probability of success in an individual trial is p, and p is constant over all trials, then the probability of x successes in n independent trials is

$$P(x) = \binom{n}{x} p^x (1-p)^{n-x}$$

To illustrate the use of the binomial law consider the following example. A firm has 10 lorries in service distributing its goods. Given that each lorry spends 10% of its time in the repair depot, what is the probability of (*a*) *no* lorry in the depot for repair, and (*b*) more than one in for repair?

(*a*) Probability of success (i.e., lorry under repair), $p = 0.10$
Number of trials $n = 10$ (lorries)

∴ Probability of no lorries being in for repair

$$P(0) = \binom{10}{0} \times 0.10^0 \times 0.90^{10} = 0.3487$$

(c.f. result obtained from first principles)

(*b*) The probability of more than one lorry being in for repair, $P(>1)$, can best be obtained by:

$$P(>1) = 1 - P(0) - P(1)$$

∴ Probability of exactly one lorry being in for repair

$$P(1) = \binom{10}{1} \times 0.10^1 \times 0.90^{10-1} = 0.3874$$

∴ Probability of more than one lorry being in for repair

$$P(>1) = 1 - 0.3487 - 0.3874 = 0.2639$$

Thus this binomial law gives a neat and quick way of computing the probabilities in simple cases like this.

1.2.11 *Management Decision Theory*

What has become known as decision theory is in simple terms just the application of probability theory and in the authors' opinion should be considered primarily as just this. This point of view will be illustrated in some examples below and in the problems given later in the chapter. It must be appreciated that in decision theory the probabilities assigned to the decisions are themselves subject to errors and, whilst better than nothing, the analysis should not be used unless a sensitivity analysis is also carried out. Also, when using decision theory (or probability theory) for decisions involving capital investment, discounted cash flow (D.C.F.) techniques are required. However, in order not to confuse readers, since this is a text on probability, D.C.F. has not been used in the examples or problems.

Note: Although it is the criterion used here by way of general introduction, the use of expected values is just one measure in a decision process. In too many books it appears to be the sole basis on which a decision is made.

Examples

1. Consider, as a simplification of the practical case, that a person wishing to sell his car has the following alternatives: (*a*) to go to a dealer with complete certainty of selling for £780, (*b*) to advertise in the press at a cost of £50, in order to sell the car for £850.

Under alternative (*b*), he estimates that the probability of selling the car for £850 is 0.60. If he does not sell through the advertisement for £850, he will take it to the dealer and sell for £780. (Note that a more realistic solution would allow for different selling prices each with their associated probability of occurrence.) Should he try for a private sale?

If he advertises the car there is a chance of 0.6 of obtaining £850 and therefore a chance of 0.4 of having to go to the dealer and accept £780.

The expected return on the sale

$$= £850 \times 0.6 + £780 \times 0.4 = £822$$

For an advertising expenditure of £50, he has only increased his expected return by £(822 − 780) or £42.

On the basis of expected return therefore, he should not advertise but go direct to the dealer and accept £780.

This method of reaching his decision is based on what would happen on average if he had a large number of cars to sell each under the same conditions as above. By advertising each of them, he would in the long run receive £8 per car less than if he sold direct to the dealer without advertising. Such a long run criterion may not be relevant to his once only decision. Compared with the guaranteed price, by advertising, he will either lose £50 or be £20 in pocket with probabilities of 0.4 and 0.6 respectively. He would probably make his decision by assessment of the risk of 40% of losing money. In practice, he could probably increase the chances of a private sale by bargaining and allowing the price to drop as low as £830 before being out of pocket.

As a further note, the validity of the estimate (usually subjective) of a 60% chance of selling privately at the price asked should be carefully examined as well as the sensitivity of any solution to errors in the magnitude of the probability estimate.

2. A firm is facing the chance of a strike occurring at one of its main plants. Considering only two points (normally more would be used), management assesses the following:

(*a*) An offer of 5% pay increase has only a 10% chance of being accepted outright. If a strike occurs:

chance of a strike lasting 1 month = 0.20
chance of a strike lasting 2 months = 0.50
chance of a strike lasting 3 months = 0.30
chance of a strike lasting longer than 3 months = 0.0

(*b*) An offer of 10% pay increase has a 90% chance of being accepted outright. If a strike occurs:

chance of strike lasting 1 month = 0.98
chance of strike lasting 2 months = 0.02
chance of strike lasting longer than 2 months = 0.0

Given that the increase in wage bill per 1% pay increase is £10 000 per month and that any agreement will last only 5 years and also that the estimated cost of a strike is £1 000 000 per month, made up of lost production, lost orders, goodwill, etc., which offer should management make?

(*a*) Considering expected costs for the *offer of 5%*. Expected loss due to strike

$$= 0.90[(0.20 \times 1) + (0.50 \times 2) + (0.30 \times 3)] \times £1\,000\,000 = £1\,890\,000$$

Increase in wage bill over 5 years

$$= £10\,000 \times 12 \times 5 \times 5 = £3\,000\,000$$

$\therefore$ Total (expected) cost of decision = £4 890 000

(*b*) For the *offer of 10%*, expected loss due to strike

$$= 0.10[(0.98 \times 1) + (0.02 \times 2)] \times £1\,000\,000 = £102\,000$$

Increase in wage bill over 5 years

= £10 000 x 12 x 5 x 10 = £6 000 000

∴ Total (expected) cost of decision = £6 102 000

Thus, management should clearly go for the lower offer and the possible strike with its consequences, although many other factors would be considered in practice before a final decision was made.

1.3 Problems for Solution

1. Four playing cards are drawn from a well-shuffled pack of 52 cards.

(*a*) What is the probability that the cards drawn will be the four aces?

(*b*) What is the probability that the cards will be the four aces drawn in order Spade, Heart, Diamond, Club?

2. Four machines–a drill, a lathe, a miller, and a grinder–operate independently of each other. Their utilisations are: drill 50%, lathe 40%, miller 70%, grinder 80%.

(*a*) What is the chance of both drill and lathe not being used at any instant of time?

(*b*) What is the chance of all machines being in use?

(*c*) What is the chance of all machines being idle?

3. A man fires shots at a target, the probability of each shot scoring a hit being 1/4 independently of the results of previous shots. What is the probability that in three successive shots

(*a*) he will fail to hit the target?

(*b*) he will hit the target at least twice?

4. Five per cent of the components in a large batch are defective. If five are taken at random and tested

(*a*) What is the probability that no defective components will appear?

(*b*) What is the probability that the test sample will contain one defective component?

(*c*) What is the probability that the test sample will contain two defective components?

5. A piece of equipment will only function if three components, *A*, *B* and *C*, are all working. The probability of *A*'s failure during one year is 5%, that of *B*'s failure is 15%, and that of *C*'s failure is 10%. What is the probability that the equipment will fail before the end of one year?

6. A certain type of seed has a 90% germination rate. If six seeds are planted, what is the chance that

(*a*) exactly five seeds will germinate?
(*b*) at least five seeds will germinate?

7. A bag contains 7 white, 3 red, and 5 black balls. Three are drawn at random without replacement. Find the probabilities that (*a*) no ball is red, (*b*) exactly one is red, (*c*) at least one is red, (*d*) all are of the same colour, (*e*) no two are of the same colour.

8. If the chance of an aircraft failing to return from any single operational flight is 5%

(*a*) what is the chance that it will survive 10 operational flights?
(*b*) if such an aircraft does survive 10 flights, what is the chance that it will survive a further 10 flights?
(*c*) if five similar aircraft fly on a mission, what is the chance that exactly two will return?

9. If the probability that any person 30 years old will be dead within a year is 0.01, find the probability that out of a group of eight such persons, (*a*) none, (*b*) exactly one, (*c*) not more than one, (*d*) at least one will be dead within a year.

10. A and B arrange to meet between 3 p.m. and 4 p.m., but that each should wait no longer than 5 min for the other. Assuming all arrival times between 3 o'clock and 4 o'clock to be equally likely, find the probability that they meet.

11. A manufacturer has to decide whether or not to produce and market a new Christmas novelty toy. If he decides to manufacture he will have to purchase a special plant and scrap it at the end of the year. If a machine costing £10 000 is bought, the fixed cost of manufacture will be £1 per unit; if he buys a machine costing £20 000 the fixed cost will be 50p per unit. The selling price will be £4.50 per unit.

Given the following probabilities of sales as:

Sales	£2000	£5000	£10 000
Probability	0.40	0.30	0.30

What is the decision with the best pay-off?

12. Three men arrange to meet one evening at the 'Swan Inn' in a certain town.

There are, however, three inns called 'The Swan' in the town. Assuming that each man is equally likely to go to any one of these inns

(*a*) what is the chance that none of the men meet?
(*b*) what is the chance that all the men meet?

13. An assembly operator is supplied continuously with components x, y, and z which are stored in three bins on the assembly bench. The quality level of the components are (1) x–10% defective, (2) y–2% defective, (3) z–5% defective.

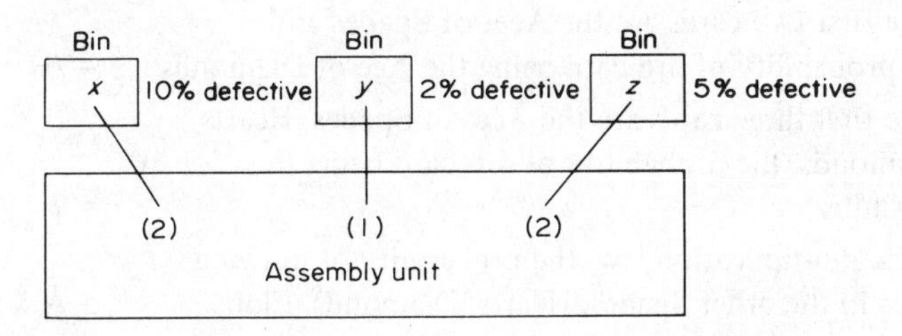

Figure 1.5

An assembly consists of two components of x, one component of y and two components of z. If components are selected randomly, what proportion of assemblies will contain

(*a*) no defective components?
(*b*) only one defective component?

14. A marketing director has just launched four new products onto the market. A market research survey showed that the chance of any given retailer adopting the products was

Product *A*	0.95	Product *C*	0.80
Product *B*	0.50	Product *D*	0.30

What proportion of retailers will (*a*) take all four new products, (*b*) take *A*, *B* and *C* but not *D*?

1.4 Solutions to Problems

1. (*a*) Probability of the 1st card being an ace $= \frac{4}{52}$

If the first card is an ace,
the probability of 2nd card being an ace $= \frac{3}{51}$

If the first two cards are aces,
the probability of 3rd card being an ace $= \frac{2}{50}$

If the first three cards are aces,
the probability of 4th card being an ace $= \frac{1}{49}$

By multiplication law,
the probability of all four being aces $= \frac{4}{52} \times \frac{3}{51} \times \frac{2}{50} \times \frac{1}{49}$
$= 0.000\,003\,7$

(*b*) Probability of 1st card being the Ace of Spades $= \frac{1}{52}$

If the first card is the Ace of Spades,
the probability of 2nd card being the Ace of Hearts $= \frac{1}{51}$

If the first two cards are the Aces of Spades and
Hearts, probability of 3rd card being the Ace of Diamonds $= \frac{1}{50}$

If the first three cards are the Aces of Spades, Hearts
and Diamonds, the probability of 4th card being the
Ace of Clubs $= \frac{1}{49}$

By the multiplication law, the probability of drawing
four aces in the order Spades, Hearts, Diamonds, Clubs $= \frac{1}{52} \times \frac{1}{51} \times \frac{1}{50} \times \frac{1}{49}$
$= 0.000\,0001\,5$

2. The utilisations can be expressed in probabilities as follows:

	Probability of being used	*Probability of being idle*
Drill	0.50	0.50
Lathe	0.40	0.60
Miller	0.70	0.30
Grinder	0.80	0.20

(*a*) By the multiplication law, the probability of drill and lathe being idle $= 0.50 \times 0.60 = 0.30$

(*b*) By the multiplication law, the probability of all machines being busy $= 0.50 \times 0.40 \times 0.70 \times 0.80 = 0.112$

(*c*) Probability of all machines being idle at any instant $= 0.5 \times 0.6 \times 0.3 \times 0.2 = 0.018$

3. (*a*) $P(\text{all three shots miss target}) = \frac{3}{4} \times \frac{3}{4} \times \frac{3}{4} = \frac{27}{64} = 0.42$

(*b*) $P(\text{hits target once}) = (\frac{1}{4} \times \frac{3}{4} \times \frac{3}{4}) + (\frac{3}{4} \times \frac{1}{4} \times \frac{3}{4}) + (\frac{3}{4} \times \frac{3}{4} \times \frac{1}{4}) = \frac{27}{64} = 0.42$

$\therefore$ $P(\text{hits target at least twice}) = 1 - (0.42 + 0.42) = 0.16$

(This result can be checked by direction evaluation of the probabilities of two hits and three hits.)

4. This problem is solved from first principles, although the binomial law can be applied.

(*a*) Probability of selecting a good item from the large batch = 0.95

By the multiplication law, probability of selecting five good items from the large batch = 0.95 x 0.95 x 0.95 x 0.95 x 0.95 = 0.77

(*b*) In a sample of five, one defective item can arise in the following five ways:

D	*A*	*A*	*A*	*A*
A	*D*	*A*	*A*	*A*
A	*A*	*D*	*A*	*A*
A	*A*	*A*	*D*	*A*
A	*A*	*A*	*A*	*D*

D = defective part
A = acceptable part

The probability of each one of these mutually exclusive ways occurring

$$= 0.05 \times 0.95 \times 0.95 \times 0.95 \times 0.95 = 0.0407$$

The probability that a sample of five will contain one defective item

$$= 5 \times 0.0407 = 0.2035$$

(*c*) In a sample of five, two defective items can occur in the following ways:

D	*D*	*D*	*D*	*A*	*A*	*A*	*A*	*A*	*A*
D	*A*	*A*	*A*	*D*	*D*	*D*	*A*	*A*	*A*
A	*D*	*A*	*A*	*D*	*A*	*A*	*D*	*D*	*A*
A	*A*	*D*	*A*	*A*	*D*	*A*	*D*	*A*	*D*
A	*A*	*A*	*D*	*A*	*A*	*D*	*A*	*D*	*D*

or in 10 ways.

Probability of each separate way = $0.05^2 \times 0.95^3$ = 0.00214

∴ Probability that the sample will contain two defectives

$$= 10 \times 0.00214 = 0.0214$$

It will be seen that permutations increase rapidly and the use of basic laws is limited. The binomial law is of course the quicker method of solving this problem, particularly if binomial tables are used.

5. The equipment would fail either if *A*, or *B*, or *C* were to fail, or if any combination of these three were to fail.

Thus the probability of the equipment failing for any reason = 1 − probability that the equipment operates for the whole year.

Probability that *A* does not fail = 0.95
Probability that *B* does not fail = 0.85
Probability that *C* does not fail = 0.90

∴ Probability that the equipment does not fail = 0.95 x 0.85 x 0.90 = 0.7268

∴ Probability that the equipment will fail = 1 − 0.7268 = 0.2732

6. (*a*) P(5 seeds germinating) = $6 \times 0.9^5 \times 0.1 = 0.3543$

(*b*) P(at least 5 seeds germinating) = P(5 or more) = P(5 or 6)

$$= P(5) + P(6) = 0.3543 + 0.9^6 = 0.8858$$

7. Conditional probability:

(*a*) Probability that no ball is red = $\frac{12}{15} \times \frac{11}{14} \times \frac{10}{13} = 0.4835$

(*b*) Probability that 1 ball is red = $3 \times (\frac{3}{15} \times \frac{12}{14} \times \frac{11}{13}) = 0.4352$

(*c*) Probability that at least 1 is red = $1 - 0.4835 = 0.5165$

(*d*) Probability that all are the same colour
= P(all white) + P(all red) + P(all black)
$= (\frac{7}{15} \times \frac{6}{14} \times \frac{5}{13}) + (\frac{3}{15} \times \frac{2}{14} \times \frac{1}{13}) + (\frac{5}{15} \times \frac{4}{14} \times \frac{3}{13}) = 0.1011$

(*e*) Probability that all are different = $6 \times \frac{7}{15} \times \frac{3}{14} \times \frac{5}{13} = 0.231$

8. (*a*) P(aircraft survives 1 flight) = 0.95
P(aircraft survives 10 flights) = $0.95^{10} = 0.7738$

(*b*) P(aircraft survives further 10 flights having survived ten)

$$= 0.95^{10} = 0.7738$$

(*c*) P(any 2 of the 5 return) = $10 \times 0.95^2 \times 0.05^3 = 0.0012$

9. (*a*) Probability that any 1 will be alive = 0.99

By the multiplication law, probability that all 8 will be alive = $0.99^8 = 0.92$

∴ Probability that none will be dead = 0.92

(*b*) By multiplication law, probability that 7 will be alive and 1 dead = $0.99^7 \times 0.01$. The number of ways this can happen is the number of permutations of 8, of which 7 are of one kind and 1 another.

$$\text{Number of ways} = \frac{8!}{7! \times 1!} = 8$$

By the addition law, probability that 7 will be alive and 1 dead

$$= 8 \times 0.99^7 \times 0.01 = 0.075$$

(*c*) By the addition law, probability of none or one being dead

$$= 0.92 + 0.075 = 0.995$$

∴ Probability of not more than one being dead = 0.995

(*d*) Probability of none being dead = 0.92

$\therefore$ Probability of 1 or more being dead = 1 – 0.92

$\therefore$ Probability of at least 1 being dead = 0.08

10. At the present stage this is best done geometrically, as in figure 1.6,

A and B will meet if the point representing their two arrival times is in the shaded area.

$$P(\text{meet}) = 1 - P(\text{point in unshaded area}) = 1 - (\tfrac{11}{12})^2 = \tfrac{23}{144}$$

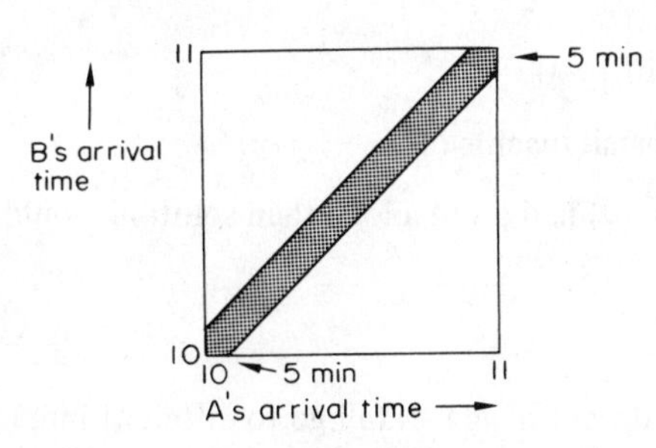

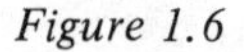
Figure 1.6

11. There are three possibilities: (*a*) to produce the toys on machine costing £10 000; (*b*) to produce the toys on machine costing £20 000; (*c*) not to produce the toys at all.

The solution is obtained by calculating the expected profits for each possibility.

(*a*) Profit on sales of 2000 = $£\left[4.50 - \left(1 + \frac{10\,000}{2000}\right)\right]$ per unit

= – £1.50 or a loss of £1.50 per unit

Profit on sales of 5000 = $£\left[4.50 - \left(1 + \frac{10\,000}{5000}\right)\right]$ per unit

= £(+ 4.50 – 3) = + £1.50

or a profit of £1.50 per unit

Profit on sales of 10 000 = $£\left[4.50 - \left(1 + \frac{10\,000}{10\,000}\right)\right]$ = + £2.50

or a profit of £2.50 per unit

$\therefore$ Expected profit = –£1.50 x 0.40 + £1.50 x 0.30 + £2.50 x 0.3

= £(–0.60 + 0.45 + 0.75) = + £0.60 per unit

(*b*) As before: profit on sales of 2000 $= £\left[4.50 - \left(0.50 + \frac{20\,000}{2000}\right)\right]$

$= -£6.00$ (i.e., loss of £6.00 per unit)

Similarly Profit on sales of 5000 $= 0$ or break even

Profit on sales of 10 000 $= +£2.00$ per unit

$\therefore$ Expected profit $= -£6.00 \times 0.4 + £0 \times 0.3 + £2.00 \times 0.3$

$= £(-2.40 + 0.60) = -£1.80$ per unit

(*c*) Expected profit = 0

$\therefore$ Solution is to install machine (*a*)

Note: If machine (*a*) had given a loss, then solution would have been not to produce at all.

12. (*a*) *P*(the men do not meet) = *P*(all go to different inns)

= *P*(1st goes to any) x *P*(2nd goes to one of the other two)

x *P*(3rd goes to last inn)

$= 1 \times \frac{2}{3} \times \frac{1}{3} = \frac{2}{9}$

(*b*) *P*(all three men meet) = *P*(1st goes to any inn)

x *P*(2nd goes to same inn)

x *P*(3rd goes to same inn)

$= 1 \times \frac{1}{3} \times \frac{1}{3} = \frac{1}{9}$

13. (*a*) There will be no defective components in the assembly if all five components selected are acceptable ones. The chance of such an occurrence is given by the product of the individual probabilities and is

$$0.90 \times 0.90 \times 0.98 \times 0.95 \times 0.95 = 0.7164$$

(*b*) If the assembly contains one defective component, any one (but only one) of the five components could be the defective. There are thus five mutually exclusive ways of getting the required result, each of these ways having its probability determined by multiplying the appropriate individual probabilities together.

1st x component	D		A		A		A		A
2nd x component	A		D		A		A		A
y component	A	or	A	or	D	or	A	or	A
1st z component	A		A		A		D		A
2nd z component	A		A		A		A		D

A = acceptable part
D = defective part

The probability of there being just one defective component in the assembly is given by

$$2 \times (0.10 \times 0.90 \times 0.98 \times 0.95 \times 0.95) + (0.90 \times 0.90 \times 0.02 \times 0.95 \times 0.95) +$$

$$+ 2 \times (0.90 \times 0.90 \times 0.98 \times 0.05 \times 0.95) = 0.1592 + 0.0146 + 0.0754 = 0.2492$$

14. Assume the products to be independent of each other. Then

(*a*) Probability of taking all four new products

$$= 0.95 \times 0.50 \times 0.80 \times 0.30 = 0.1140$$

(*b*) Probability of taking only

$$A, B, \text{and } C = 0.95 \times 0.50 \times 0.80 \times (1 - 0.30) = 0.2660$$

1.5 Practical Laboratory Experiments and Demonstrations

Experience has shown that when students are being introduced to statistics, the effectiveness of the course is greatly improved by augmenting it with a practical laboratory course of experiments and demonstrations, irrespective of the mathematical background of the students.

The three experiments described here are experiments 1, 2, and 3 from the authors' *Laboratory Manual in Basic Statistics,* which contains full details, analysis and summary sheets.

Appendix 1 gives full details of experiment 1 together with the analysis and summary sheets.

The following notes are for guidance on experiments.

1.5.1 *Experiment 1*

This experiment, in being the most comprehensive of the experiments in the book, is unfortunately also the longest as far as data collection goes. However, as will be seen from the points made, the results more than justify the time. Should time be critical it is possible to miss experiment 1 and carry out experiments 2 and 3 which are much speedier. In experiment 1 the data collection time is relatively long since the three dice have to be thrown 100 times (this cannot be reduced without drastically affecting the results).

Appendix 1 contains full details of the analysis of eight groups' results for the first experiment, and the following points should be observed in summarising the experiment:

(1) The variation between the frequency distributions of number of *ones* (or number of *sixes*) obtained by all groups, and that the distributions based on total data (sum of all groups) are closer to the theoretical situation.

(2) The comparison of the distributions of score of the coloured dice and the total score of three dice show clearly that the total score distribution now tends to a bell-shaped curve.

1.5.2 *Experiment 2*

This gives a speedy demonstration of Bernoulli's law. As n, the number of trials, increases, the estimate of p the probability gets closer to the true population value. For $n = 1$ the estimate is either $p = 1$ or 0 and as n increases, the estimates tend to get closer to $p = 0.5$. Figure 1.7 shows a typical result.

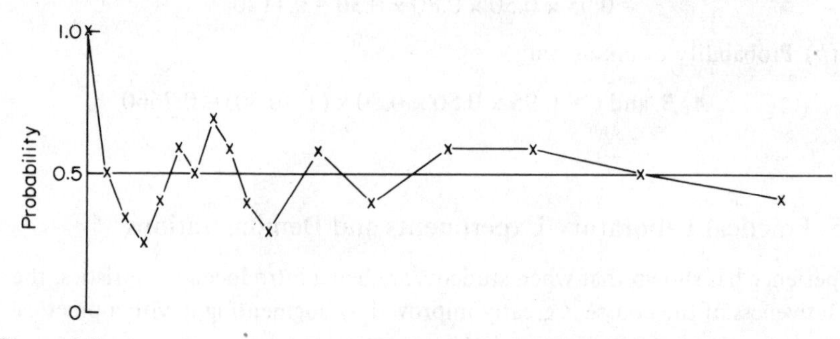

Figure 1.7

1.5.3 *Experiment 3*

Again this is a simple demonstration of probability laws and sampling errors. Four coins are tossed 50 times and in each toss the number of heads is recorded. See table 6 of the laboratory manual.

Note

It is advisable to use the specially designed shakers or something similar. Otherwise the coins will roll or bias in the tossing will occur. The results of this experiment are summarised in table 8 of the laboratory manual and the variation in groups' results are stressed as is the fact that the results based on all groups' readings are closer to the theoretical than those for one group only.

1.5.4 *Summary of Experiments 1, 2, and 3*

The carrying out of these experiments will have given students a feel for the basic concepts of statistics. While in all other sciences they expect their results to obey the theoretical law exactly, they will have been shown that in statistics all samples vary, but an underlying pattern emerges. The larger the samples used the closer this pattern tends to be to results predicted by theory. The basic laws–those of addition and multiplication–and other concepts of probability theory, have been illustrated.

Other experiments with decimal dice can be designed.†

Appendix 1–Experiment 1 and Sample Results

Probability Theory

Number of persons: 2 or 3.

Object

The experiment is designed to illustrate

(*a*) the basic laws of probability

(*b*) that the relative frequency measure of probability becomes more reliable the greater the number of observations on which it is based, that is, Bernoulli's theorem.

Method

Throw three dice (2 white, 1 coloured) a hundred times. For each throw, record in table 1

(*a*) the number of *ones*
(*b*) the number of *sixes*
(*c*) the score of the coloured die
(*d*) the total score of the three dice.

Draw up these results, together with those of other groups, into tables (2, 3, and 4).

Analysis

1. For each set of 100 results and for the combined figures of all groups, calculate the probabilities that, in a throw of three dice:

(*a*) no face shows a *one*
(*b*) two or more *sixes* occur
(*c*) a total score of *more than* 13 is obtained

† Details from: Technical Prototypes (Sales) Limited, 1A West Holme Street, Leicester.

Table 1

No. of 1's	No. of 6's	Coloured die	Total score	No. of 1's	No. of 6's	Coloured die	Total score	No. of 1's	No. of 6's	Coloured die	Total score	No. of 1's	No. of 6's	Coloured die	Total score
0	1	5	16	0	1	5	13	0	0	2	8	0	0	3	11
1	0	2	8	2	0	1	4	0	0	4	11	0	0	2	10
0	0	4	14	0	0	5	14	0	1	6	12	1	1	1	9
1	0	5	8	0	0	2	7	1	1	1	10	0	1	2	11
2	0	4	6	1	0	4	10	1	0	5	11	1	0	1	9
0	1	6	16	1	1	1	12	0	0	3	8	0	1	6	12
0	1	5	15	0	0	4	13	0	0	2	9	0	0	3	7
0	0	5	9	0	1	4	13	0	0	2	6	1	1	1	12
2	0	1	5	1	0	1	6	0	1	4	13	0	0	5	13
0	0	3	12	1	0	1	10	0	1	2	12	2	1	1	8
2	0	1	4	0	1	6	12	0	1	6	15	0	0	5	12
1	0	3	7	0	2	6	16	0	2	6	16	0	0	2	9
0	1	3	14	0	1	6	13	1	0	1	8	0	1	4	15
2	0	2	4	1	0	1	6	0	0	5	13	0	1	3	13
0	2	3	15	0	1	6	12	0	0	2	8	0	1	5	16
0	0	2	7	1	0	5	11	1	0	1	10	0	1	3	13
0	1	6	12	1	1	6	10	0	0	5	13	0	1	4	12
2	0	2	4	1	0	5	9	0	0	3	11	0	1	4	14
0	1	6	12	0	0	3	9	0	1	4	13	0	0	5	13
2	0	1	7	0	0	5	10	0	1	5	13	0	0	4	10
0	0	3	9	1	0	5	9	1	2	1	13	0	0	5	10
1	0	1	6	1	1	6	10	0	2	6	14	1	1	1	12
2	0	1	4	2	0	1	7	1	1	3	10	2	1	1	8
1	1	1	10	1	0	1	7	1	1	1	12	0	1	6	14
0	0	3	7	0	0	4	12	0	1	6	15	2	0	1	4

Table 2

Tally marks	Total score	Frequency of given total score for each group								Total freq^y for all groups	Experi-mental proba-bility	Theo-retical proba-bility
		1	②	3	4	5	6	7	8			
—	3	0	0	0	0	0	0	0	2	2	0·0025	0.0046
𝍸 I ✓	4	0	6	0	0	1	1	4	0	12	0·015	0.0138
I ✓	5	3	1	4	1	1	2	2	4	18	0·0225	0.0278
𝍸 ✓	6	4	5	6	2	8	10	4	3	42	0·0525	0.0463
𝍸 III ✓	7	9	8	7	5	11	6	7	9	62	0·0775	0.0694
𝍸 III ✓	8	16	8	10	17	12	8	17	10	98	0·1225	0.0972
𝍸 IIII ✓	9	12	9	8	10	13	8	4	6	70	0·0875	0.1157
𝍸 𝍸 II ✓	10	17	12	14	16	12	12	12	13	108	0·135	0.1250
𝍸 I ✓	11	7	6	15	14	9	6	13	19	99	0·1235	0.1250
𝍸 𝍸 𝍸 ✓	12	12	15	10	10	10	16	9	10	92	0·115	0.1157
𝍸 𝍸 IIII ✓	13	9	14	9	10	7	12	9	8	78	0·0975	0.0972
𝍸 I ✓	14	4	6	5	5	6	5	9	5	45	0·0562	0.0694
𝍸 ✓	15	3	5	8	4	4	1	4	6	35	0·0437	0.0463
𝍸 ✓	16	3	5	3	5	4	2	4	3	29	0·0363	0.0278
—	17	1	0	0	0	1	1	2	1	6	0·0075	0.0138
—	18	0	0	1	1	1	0	0	1	4	0·005	0.0046
No. of throws		100	100	100	100	100	100	100	100	800		
Probability of score of more than 13		0·11	0·16	0·17	0·15	0·16	0·09	0·19	0·16			

2. Compare these results with those expected from theory and comment on the agreement both for individual groups and for the combined observations.

3. Draw probability histograms both for the score of the coloured die and for the total score of the three dice, on page 27. Do this for your own group's readings and for the combined results of all groups.

Comment on the agreement with the theoretical distributions.

Note: The theoretical probability distribution for the total score of three dice is shown in table 2.

Table 3

Group	No. of throws	No. of throws in which given no. of ONES occur				Probability that, in one throw, no face shows a ONE	No. of throws in which given no. of SIXES occur				Probability that, in one throw, two or more SIXES appear
		0	1	2	3		0	1	2	3	
1		62	40	8	0	0·52	62	32	6	0	0·06
2	100	61	27	12	0	0·61	55	40	5	0	0·05
3		54	38	8	0	0·54	58	37	4	1	0·05
4		56	41	3	0	0·56	59	34	6	1	0·03
5		55	40	5	0	0·55	63	32	4	1	0·05
6		65	26	9	0	0·65	61	36	3	0	0·03
7		57	35	8	0	0·57	59	34	7	0	0·09
8		56	39	3	2	0·56	59	35	5	1	0·06
Totals	800	456	286	56	2		474	280	42	4	
Experimental probability		0·57	0·357	0·07	0·0025		0·5925	0·35	0·0525	0·005	
Theoretical probability		0·578	0·347	0·0695	0·0046	0·578	0·578	0·347	0·0695	0·0046	0·074

Table 4

Group	No. of throws	No. of throws in which given score appears on the coloured die						Probability of odd number
		1	2	3	4	5	6	
1	↑	15	20	24	12	18	11	0·57
②	100	25	13	14	13	19	16	0·58
3		18	20	29	14	14	15	0·51
4		15	15	15	14	20	21	0·50
5		20	16	22	16	17	9	0·59
6		15	18	15	16	21	15	0·51
7		18	19	13	18	17	15	0·58
8	↓	11	17	23	14	18	17	0·52
Totals	800	137	138	145	117	144	119	
Experimental probability		0·171	0·173	0·181	0·146	0·180	0·149	
Theoretical probability		0·167	0·167	0·167	0·167	0·167	0·167	0·50

2 Theory of distributions

2.1 Syllabus Covered

Summary of data; frequency and probability distributions; histograms; samples and populations; distribution types; moments and their calculation; suggested experiments and demonstrations.

2.2 Résumé of Basic Theory and Concepts

2.2.1 *Introduction*

The understanding of the concepts of distributions and their laws is fundamental to the science of statistics. Variation occurs almost without exception in all our activities and processes. For example nature cannot produce two of her products alike—two 'identical' twins are never exactly alike; a description of similarity is the saying 'as alike as two peas', yet study two peas from the same pod and differences in size or colour or shape will become apparent. Consider for example the heights of men. Heights between 1.70 m and 1.83 m are quite common and heights outside this range are by no means rare.

Although it is not so obvious, man-made articles are also subject to the same kind of variability. The manufacturer of washers realises that some washers will have a greater thickness than others. The resistance of electrical filaments made at the same time will not be exactly alike. The running cost of a department in a company will not be exactly the same each week, although, off hand, there is no reason for the difference. The tensile strength of a steel bar is not the same at two different parts of the same bar. The ash content of coal in a truck is different when a number of samples from the truck are tested. Differences in the diameter of components being produced on the same lathe will be found. The time taken to do a given job will vary from occasion to occasion.

In present-day manufacture, the aim is usually to make things as alike as possible. Or, alternatively, the amount of variability is controlled by specification so that variation between certain limits is permitted.

It is interesting to note that even with the greatest precision of manufacture, variability will still exist, providing the measuring equipment is sensitive enough to pick it up.

Thus, variation will be seen to be present in all processes, to a greater or lesser extent, and the use of distributions and their related theorems is necessary to analyse such situations.

2.2.2 *Basic Theory of Distributions*

The basic concepts of distributions can be illustrated by considering any collection of data such as the 95 values of the output time for an open hearth furnace given in table 2.1. The output time is the overall time from starting to charge to finishing tapping the furnace.

7.8	8.0	8.6	8.1	7.9	8.2	8.1	7.9	8.2	8.1
8.4	8.2	7.8	8.0	7.5	7.4	8.0	7.3	7.6	7.8
7.7	7.8	7.5	7.9	7.8	8.3	7.9	8.0	8.2	7.4
7.1	7.5	7.9	8.2	8.5	7.9	7.5	7.8	8.4	8.1
8.2	7.9	8.7	7.7	7.8	8.0	8.1	8.2	7.9	7.3
8.0	8.1	7.8	8.1	7.6	7.8	7.9	8.5	7.8	
8.3	7.9	8.1	7.6	7.9	8.3	7.4	7.9	8.7	
7.6	8.0	8.0	8.2	8.2	7.9	8.1	8.4	7.6	
7.9	7.7	7.9	7.8	7.8	7.7	7.5	7.7	8.1	
8.1	8.0	8.1	7.7	8.0	8.0	8.0	8.1	7.7	

Table 2.1 Furnace output time (h)

Referring to these data, it will be seen that the figures *vary* one from the other; the first is 7.8 h, the next 8.4 h and so on; there is one as low as 7.1 h and one as high as 8.7 h.

In statistics the basic logic is inductive, and the data must be looked at as a whole and not as a collection of individual readings.

It is often surprising to the non-statistician or deterministic scientist how often regularities appear in these statistical counts.

The process of grouping data consists of two steps usually carried out together.

Step 1. The data are placed in order of magnitude.
Step 2. The data are then summarised into groups or class intervals.

This process is carried out as follows:

(1) The range of the data is found, i.e.

$$\text{largest reading} - \text{smallest reading} = 8.7 - 7.1 = 1.6 \text{ h}$$

(2) The range is then sub-divided into a series of steps called *class intervals.*

These class intervals are usually of equal size, although in certain cases unequal class intervals are used. For usual sample sizes, the number of class intervals is chosen to be between 8 and 20, although this should be regarded as a general rule only. For table 2.1, class intervals of size 0.2 h were chosen, i.e.,

7.1–7.3, 7.3–7.5, . . ., 8.7–8.9

(3) More precise definition of the boundaries of the class intervals is however required, otherwise readings which fall say at 7.3 can be placed in either of two class intervals.

Since in practice the reading recorded as 7.3 h could have any value between 7.25 h and 7.35 h (normal technique of rounding off), the class boundaries will now be taken as:

7.05–7.25, 7.25–7.45, . . ., 8.45–8.65, 8.65–8.85

Note: Since an extra digit is used there is no possibility of any reading's falling on the boundary of a class.

The summarising of data in figure 2.1 into a distribution is shown in table 2.2. For each observation in table 2.1 a stroke is put opposite the sub-range into which the reading falls. The strokes are made in groups of five for easy summation.

Value of variable	Frequency	Frequency distribution	Probability distribution
7.05–7.25	I	1	0.01
7.25–7.45	IIII	5	0.05
7.45–7.65	IIII IIII	10	0.11
7.65–7.85	IIII IIII IIII IIII	19	0.20
7.85–8.05	IIII IIII IIII IIII IIII II	27	0.28
8.05–8.25	IIII IIII IIII IIII II	22	0.23
8.25–8.45	IIII I	6	0.06
8.45–8.65	III	3	0.03
8.65–8.85	II	2	0.02
		Total = 95	Total = 1.00

Table 2.2

The last operation is to total the strokes and enter the totals in the next to last column in table 2.2 obtaining what is called a frequency distribution. There are for example, one reading in class interval 7.05–7.25, five readings in the next, ten in the next, and so on. Such a table is called a frequency distribution since it shows how the individuals are distributed between the groups or class intervals. Diagrams are more easily assimilated so it is normal to plot the

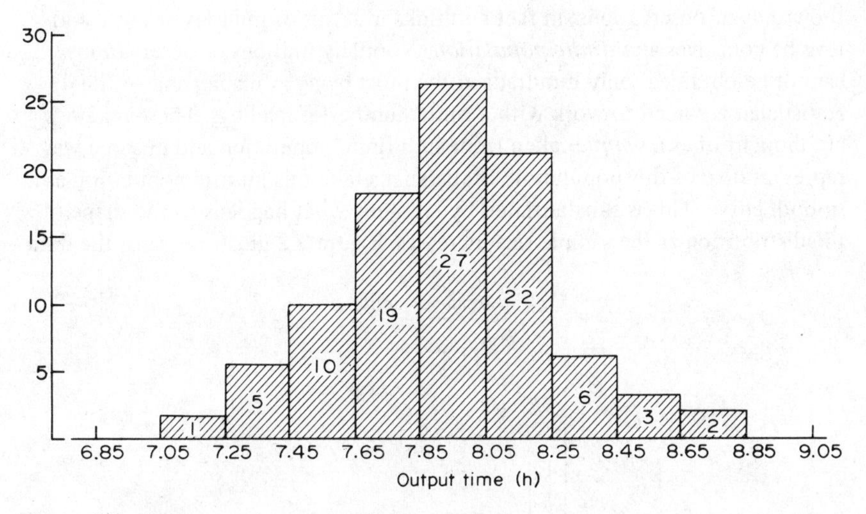

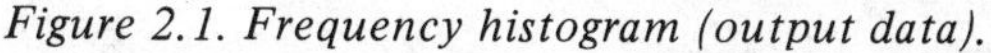
Figure 2.1. Frequency histogram (output data).

frequency distribution and this frequency histogram is shown in figure 2.1.

In plotting a histogram, a rectangle is erected on each class interval, the area of the rectangle being proportional to the class frequency.

Note: Where class intervals are all of equal length, the height of the rectangle is also proportional to the class frequency. Other examples of frequency distributions and histograms are given in section 2.3.

2.2.3 *Probability Distributions*

The frequency distribution is often transformed into a probability distribution by calculating the relative frequency or probability of a reading falling in each class interval.

For example, probability of a reading falling in the interval 7.45–7.65

$$= \frac{\text{number of readings in class}}{\text{total number of readings}} = \frac{10}{95} = 0.11$$

(See chapter 1 on measurement of probability.)

Probability distributions have a distinct advantage when comparing two or more sets of data since the area under the curve has been standardised in all cases to unity.

2.2.4 *Concept of a Population*

All the data of table 2.1 are summarised by means of the frequency distribution shown in figure 2.1. The distribution was obtained from a sample of 95 observations. However, in statistics the analyst likes to think in terms of thousands of observations; in fact he thinks in terms of millions or more and thus he conceives an *infinite population.* Normally millions of observations cannot be obtained, only hundreds at the most being available, and so the statistician is forced to work with a finite number of readings. These readings are thought of as a *sample* taken from an infinite population and in some way representative of this population. Statisticians take this infinite population as a smooth curve. This is substantiated by studying what happens to the shape of the distribution as the sample size increases. Figure 2.2 illustrates this, the data

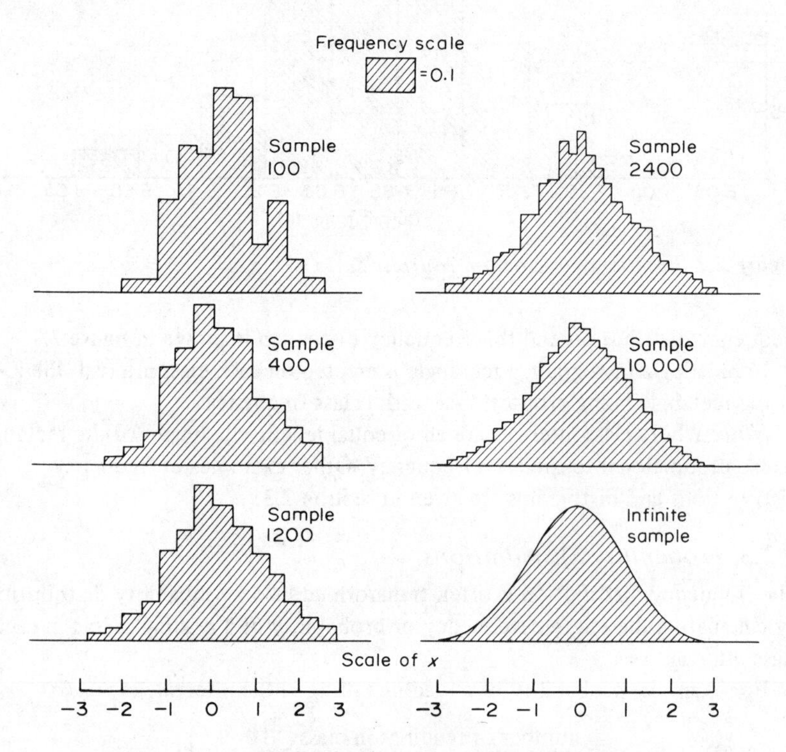

Figure 2.2. The effect of the sample size on the histogram shape.

here being taken from an experiment in a laboratory. A sample size of 100 gives an irregular shape similar to those obtained from the data of output times. However, with increasing sample size, narrower class intervals can be used and the frequency distribution becomes more uniform in shape until with a sample of 10 000 it is almost smooth. The limit as the sample size becomes infinite is also shown. Thus with small samples, irregularities are to be expected in the frequency distributions, even when the population gives a smooth curve.

It is the assumption that the population from which the data was obtained has a smooth curve (although not all samples have), that enables the statistician to use the mathematics of statistics.

2.2.5 *Moments of Distribution*

The summarising of data into a distribution always forms the first stage in statistical analysis. However, this summarising process must usually be taken further since a shape is not easy to deal with.

The statistician, in the final summary stage, calculates measures from the distribution, these measures being used to represent the distribution and thus the original data.

Each measure is called a *statistic.* In calculating these measures or statistics, the concept of *moments* is borrowed from mechanics. The distribution in probability form is considered as lying on the x axis and the readings in each interval as having the value of the mid-point of each interval, i.e., $x_1, x_2, \ldots, x_N$ etc.

If the probabilities associated with these variable values are $p_1, p_2, \ldots p_N$ (figure 2.3 shows this diagrammatically), then $p_1 + p_2 + \ldots + p_N = 1$.

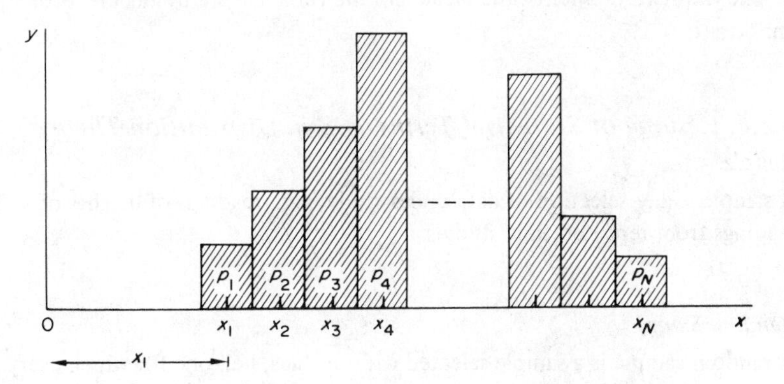

Figure 2.3

Consider now the 1st moment of the distribution about the origin

$$= \sum_{i=1}^{N} p_i x_i = \bar{x} \text{ (the arithmetical average)}$$

Thus the 1st *statistic* or *measure* is the arithmetical average $\bar{x}$. Higher moments are now taken about this arithmetical average rather than the origin.

Thus, the 2nd moment about the arithmetical average

$$= \sum_{i=1}^{N} p_i(x_i - \bar{x})^2$$

This 2nd moment is called the *variance* in statistics, and its square root is called the *standard deviation.*

Thus the standard deviation of the distribution

$$= \sqrt{\left[\sum_{i=1}^{N} p_i(x_i - \bar{x})^2\right]}$$

The higher moments are as follows:

$$\text{3rd moment about the average} = \sum_{i=1}^{N} p_i(x_i - \bar{x})^3$$

$$\text{4th moment about the average} = \sum_{i=1}^{N} p_i(x_i - \bar{x})^4$$

$$\text{or in general the } k\text{th moment about the average} = \sum_{i=1}^{N} p_i(x_i - \bar{x})^k$$

The first two moments, the mean and the variance, are by far the most important.

2.2.6 *Résumé of Statistical Terms used in Distribution Theory*

Sample

A sample is any selection of data under study, e.g., readings of heights of men, readings from repeated time studies.

Random Sample

A random sample is a sample selected without bias, i.e., one for which every member of the population has an equal chance of being included in the sample.

Population or Universe

This is the total number of possible observations. This concept of a population is fundamental to statistics. All data studied are in sample form and the statistician's sample is regarded as having been drawn from the population of all possible events. A population may be finite or infinite. In practice, many finite populations are so large they can be conveniently considered as infinite in size.

Grouping or Classification of Numerical Data

The results are sub-divided into groups so that no regard is paid to variations within the groups. The following example illustrates this.

Groupings	*Number of results*
⋮	⋮
3.95–4.95	8
4.95–5.95	7
5.95–6.95	5
⋮	⋮

The class boundaries shown in this example are suitable for measurements recorded to the nearest 0.1 of a unit. The boundaries chosen are convenient for easy summary of the raw data since the first class shown contains all measurements whose integer part is 4, the next class all measurements starting with 5 and so on.

It would have been valid but less convenient to choose the class as, say, 3.25–4.25, 4.25–5.25, . . .

In grouping, any group is called a *class* and the number of values falling in the class is the *class* frequency. The magnitude of the range of the group is called the class interval, i.e., 3.95–4.95 or 1.

Number of Groups

For simplicity of calculation, the number of intervals chosen should not be too large, preferably not more than twenty. Again, in order that the results obtained may be sufficiently accurate, the number must not be too small, preferably not less than eight.

Types of Variable

Continuous. A continuous variable is one in which the variable can take every value between certain limits a and b, say.

Discrete. A discrete variable is one which takes certain values only—frequently part or all of the set of positive integers. For example, each member of a sample may or may not possess a certain attribute and the observation recorded (the value of the variable) might be the number of sample members which possess the given attribute.

Frequency Histogram

A frequency distribution shows the number of samples falling into each class interval when a sample is grouped according to the magnitude of the values. If the class form, frequency is plotted as a rectangular block on the class interval the diagram is called a frequency histogram. *Note:* Area is proportional to frequency.

Probability Histograms

A probability histogram is the graphical picture obtained when the grouped sample data are plotted, the class probability being erected as a rectangular block on the class interval. The area above any class interval is equal to the probability of an observation being in that class since the total area under the histogram is equal to *one.*

Limiting form of Histogram

The larger the sample, the closer the properties of histograms and probability curves become to those of the populations from which they were drawn, i.e., the limiting form.

Variate

A variate is a variable which possesses a probability distribution.

2.2.7 *Types of Distribution*

While there is much discussion as to the value of classifying distributions into types, there is no doubt in the authors' minds that classification does help the student to get a better appreciation of the patterns of variation met in practice.

Figure 2.4 gives the usually accepted classifications.

Type 1: Unimodal

Examples of this variation pattern are: intelligence quotients of children, heights (and/or weights) of people, nearly all man-made objects when produced under controlled conditions (length of bolts mass-produced on capstans, etc.).

A simple example of this type of distribution can be illustrated if one assumes that the aim is to make each item or product alike but that there exists a very large number of small independent forces deflecting the aim, and under such conditions, a unimodal distribution arises. For example, consider a machine tool mass-producing screws. The setter sets the machine up as correctly as he can and then passes it over to the operator and the screws produced form a pattern of variation of type 1. The machine is set to produce each screw exactly the same, but, because of a large number of deflecting forces present, such as small particles of grit in the cooling oil, vibrations in the machine, slight variation in the metal–manufacturing conditions are not constant, hence there is variation in the final product. (See simple quincunx unit on page 61.)

Type 2: Positive Skew

Examples of this type of distribution are the human reaction time and other types of variable where there is a lower limit to the values, i.e., distribution of number of packages bought at a supermarket, etc.

If this type of distribution is met when a symmetrical type should be expected it is indicative of the process being out of control.

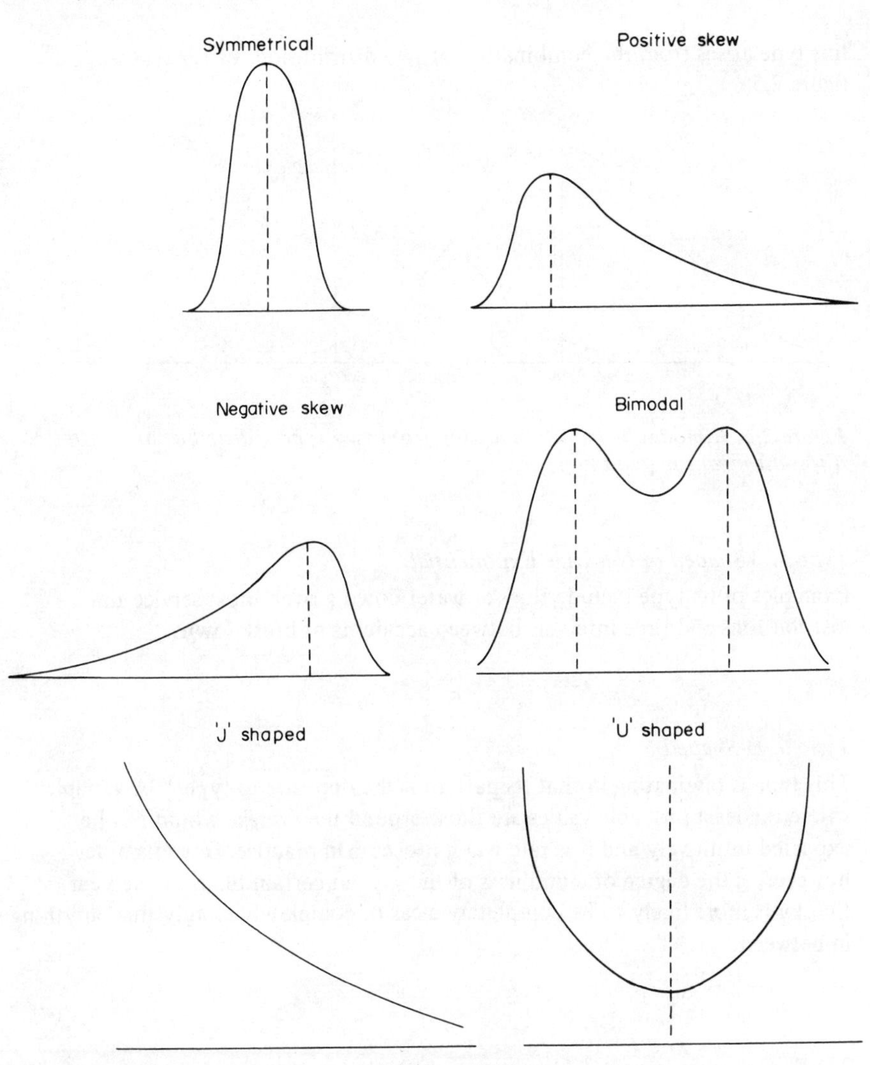

Figure 2.4. Types of distribution.

Type 3: Negative Skew

True examples of this type are difficult to find in practice but can arise when there is some physical or other upper constraint on the process.

Type 4: Bimodal

This type cannot be classified as a separate form unless more evidence of measures conforming to this pattern of variation are discovered. In most cases this type arises from the combination of two distributions of type 1 (see figure 2.5).

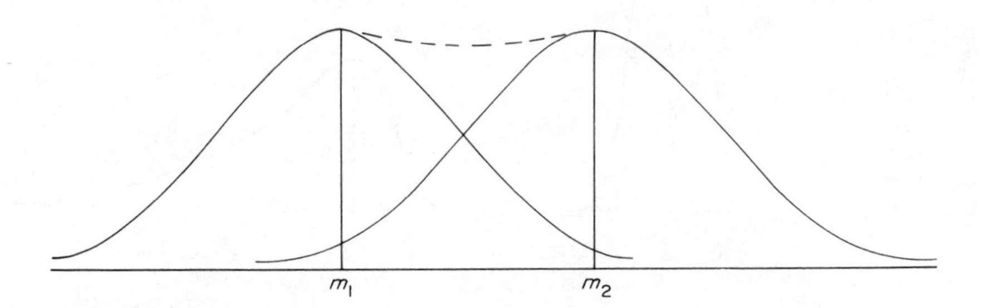

Figure 2.5. Bimodal distribution arising from two type-1 distributions with different means m_1 and m_2.

Type 5: J-Shaped or Negative Exponential

Examples of its type include flow of water down a river, most service time distributions and time intervals between accidents or breakdowns.

Type 6: U-Shaped

This type is fascinating in that its pattern is the opposite to type 1. A variable where the least probable values are those around the average would not be expected intuitively and it is rare when it occurs in practice. One example, however, is the degree of cloudiness of the sky—at certain times of the year the sky is more likely to be completely clear or completely cloudy than anything in between.

2.2.8 *Computation of Moments of Distribution*

Dependent on the type of data and their range, the data may or may not be grouped into class intervals. The calculation of moments is the same for either grouped or ungrouped data, but in the case of grouped data, all the readings in the class interval are regarded as lying at the centre of the interval. The method used in this text and throughout all the examples makes use of a simple transformation of the variate and is usually carried out on the frequency distributions rather than on the probability distribution. This use of frequency distributions is common to most text books and will be used here although there is often advantage in using the probability distribution.

Let x_i = value of the variate in the ith class interval
f_i = frequency of readings in the ith class interval
p_i = probability of a value in the ith class interval
$\sum_i f_i = n$, the total number of readings

$$\text{The 1st moment (arithmetic average)} = \frac{\sum_i f_i x_i}{\sum_i f_i} = \bar{x}$$

or

$$\sum_i p_i x_i = \bar{x}$$

The 2nd moment (variance)
$$(s')^2 = \frac{\sum f_i (x_i - \bar{x})^2}{\sum f_i}$$

or

$$\sum_i p_i (x_i - \bar{x})^2$$

For computing purposes the formula for variance is usually modified to reduce the effect of rounding errors. These errors can arise through use of the calculated average $\bar{x}$ which is generally a rounded number. If insufficient significant figures are retained in $\bar{x}$, each of the deviations $(x_i - \bar{x})$ will be in error and the sum of their squares $[\Sigma f_i (x_i - \bar{x})^2]$ will tend to be inaccurate.

Computation of Moments using Frequency Distributions

The variate (x_i) is transformed to

$$u_i = \frac{x_i - x_0}{c} \quad \text{or} \quad x_i = cu + x_0$$

where x_0 = any value of x taken as an arbitrary average, c = class interval width.

It can easily be shown that

$$\text{1st moment } \bar{x} = x_0 + \frac{c \sum_i f_i u_i}{\sum_i f_i}$$

$$\text{2nd moment } (s')^2 = c^2 \frac{\left[\Sigma f_i u_i^2 - \frac{(\Sigma f_i u_i)^2}{\Sigma f_i}\right]}{\Sigma f_i}$$

Example

The values given in table 2.3 have been calculated using the data from table 2.2.

Class interval	Mid point (x_i)	Frequency (f_i)	u_i	$f_i u_i$	$f_i u_i^2$
7.05–7.25	7.15	1	−4	−4	16
7.25–7.45	7.35	5	−3	−15	45
7.45–7.65	7.55	10	−2	−20	40
7.65–7.85	7.75	19	−1	−19	19
7.85–8.05	7.95	27	0	0	0
8.05–8.25	8.15	22	+1	+22	22
8.25–8.45	8.35	6	+2	+12	24
8.45–8.65	8.55	3	+3	+9	29
8.65–8.85	8.75	2	+4	+8	32
		$\Sigma f_i = 95$		$\Sigma f_i u_i = -7$	$\Sigma f_i u_i^2 = 225$

Table 2.3

Let $x_0 = 7.95$, $c = 0.20$ h

$$\therefore \quad \text{arithmetic average} = 7.95 + 0.20\left(\frac{-7}{95}\right) = 7.94 \text{ h}$$

$$\therefore \quad \text{variance} \quad (s')^2 = (0.2)^2 \left[\frac{225 - \left(\frac{-7}{95}\right)^2}{95}\right] = 0.095$$

Computation using Probability Distributions

Class interval	Mid point (x_i)	Probability (p_i)	u_i	$u_i p_i$	$u_i^2 p_i$
7.05–7.25	7.15	0.01	−4	−0.04	0.16
7.25–7.45	7.35	0.05	−3	−0.15	0.45
7.45–7.65	7.55	0.11	−2	−0.22	0.44
7.65–7.85	7.75	0.20	−1	−0.20	0.20
7.85–8.05	7.95	0.28	0	0	0
8.05–8.25	8.15	0.23	+1	+0.23	0.23
8.25–8.45	8.35	0.06	+2	+0.12	0.24
8.45–8.65	8.55	0.03	+3	+0.09	0.27
8.65–8.85	8.75	0.02	+4	+0.08	0.32

Table 2.4

Let $x_0 = 7.95$ and $c = 0.2$ then $\Sigma u_i p_i = 0.09$ and $\Sigma(p_i u_i^2) = 2.31$

The formulae for the moments are

arithmetic average $\bar{x} = x_0 + c\,\Sigma p_i u_i = 7.95 + (-0.018) = 7.93$ h

variance $(s')^2 = c^2\,[\Sigma p_i u_i^2 - (\Sigma p_i u_i)^2] = 0.2^2\,(2.31 - 0.09^2) = 0.092$

which compares favourably with results achieved using the frequency distribution in view of the rounding off of probability to the second decimal point.

2.2.9 *Sheppard's Correction*

When calculating the moments of grouped distributions, the assumption that the readings are all located at the centre of the class interval leads to minor errors in these moments. It must be stressed that the authors do not consider that these corrections, known as Sheppard's corrections, are of sufficient magnitude in most problems, to be used.

However, it is only correct that they should be given:

$$\text{Correction to 1st moment, } \bar{x} = 0$$

$$\text{Correction to 2nd moment } = -\frac{c^2}{12}$$

Thus the 1st moment calculation is unbiased while the answer given for the 2nd moment should be reduced by $c^2/12$.

2.3 Problems for Solution

In the following problems, students are required to

(1) summarise data into distributions
(2) draw the frequency histogram
(3) calculate the mean and standard deviation of data.

While there is a large collection of problems given, tutors should select those examples most relevant to their students' courses. Worked solutions are given for all questions in section 2.4, but in the authors' opinion the answering of two or three problems should be adequate.

Note: Students' answers may differ slightly from the given answers, depending on the class intervals selected.

The distributions illustrate that with limited samples of 30 to 100 observations the shapes of the distributions can tend in some cases to be relatively irregular.

1. In a work study investigation of an operator glueing labels onto square biscuit tins, the following readings, in basic minutes, were obtained for the time of each operation:

0.09	0.09	0.11	0.09	0.09	0.11	0.09	0.07	0.09	0.06
0.09	0.09	0.09	0.11	0.09	0.07	0.09	0.06	0.10	0.07
0.09	0.10	0.06	0.10	0.08	0.06	0.09	0.08	0.08	0.08
0.08	0.10	0.08	0.07	0.09	0.08	0.09	0.11	0.09	0.09
0.08	0.10	0.09	0.08	0.10	0.08	0.08	0.09	0.09	0.09
0.08	0.06	0.08	0.08	0.10	0.09	0.09	0.10	0.10	0.11

2. In the assembly of a Hoover agitator, time element number 2 consists of: pick up two spring washers, one in each hand and place on spindle, pick up two bearings and place on spindle, pick up two felt washers and place on spindle, pick up two end caps and screw onto spindle.

The following data, in basic minutes, were obtained from 93 studies for the time element number 2:

0.26	0.28	0.31	0.22	0.25	0.28	0.28	0.26	0.29	0.25
0.24	0.29	0.26	0.28	0.24	0.26	0.29	0.23	0.26	0.26
0.25	0.30	0.25	0.29	0.17	0.26	0.33	0.24	0.18	0.34
0.26	0.31	0.23	0.29	0.22	0.26	0.29	0.25	0.24	0.28
0.27	0.32	0.23	0.26	0.25	0.28	0.36	0.42	0.24	0.21
0.23	0.27	0.46	0.23	0.28	0.31	0.29	0.31	0.25	
0.24	0.28	0.33	0.24	0.29	0.36	0.32	0.27	0.24	
0.25	0.29	0.33	0.25	0.35	0.24	0.33	0.28	0.26	
0.26	0.20	0.24	0.26	0.34	0.30	0.30	0.29		
0.18	0.22	0.25	0.27	0.33	0.30	0.30	0.23		

3. The time interval, in minutes, between the arrival of successive customers at a cash desk of a self-service store was measured over 56 customers and the results are given below:

1.05	1.68	0.78	1.10	0.32	1.61	0.10	0.43	3.70	0.09
0.21	2.71	2.12	2.81	3.30	0.15	0.54	3.12	0.80	1.76
1.14	0.16	0.31	0.91	0.18	0.04	1.16	2.16	1.48	0.63
0.57	0.65	4.60	1.72	0.52	2.32	0.08	0.62	3.80	1.21
1.16	0.58	0.57	0.04	1.19	0.11	0.05	2.68	2.08	0.01
0.15	0.42	0.25	0.05	1.88	3.90				

4. The number of defects per shift from a large indexing machine are given below for the last 52 shifts:

2 6 4 5 1 3 2 1 4 2 1 4 6
3 4 3 2 4 5 4 3 6 3 0 7 4
7 3 5 4 3 2 0 5 2 5 3 2 9
5 3 2 1 1 0 3 3 1 4 1 3 2

5. The crane handling times, in minutes, for a sample of 100 jobs lifted and moved by an outside yard mobile crane are given below:

5 6 21 8 7 8 11 5 10 21
13 15 17 7 27 6 6 11 9 4
7 4 9 192 10 15 31 15 11 38
16 52 87 20 18 22 11 7 9 8
6 10 10 17 37 32 10 26 14 15
28 182 17 27 4 9 19 10 44 20
15 5 20 8 25 14 23 13 12 7
9 92 33 22 19 151 171 21 4 6
31 13 7 45 6 7 17 7 19 42
9 6 55 61 52 7 5 102 8 23

6. The lifetime, in hours, of a sample of 100 electric light bulbs is given below:

1067 919 1196 785 1126 936 918 1156 920 1192
855 1092 1162 1170 929 950 905 972 1035 922
1022 978 832 1009 1157 1151 1009 765 958 1039
923 1333 811 1217 1085 896 958 1311 1037 1083
999 932 1035 944 1049 940 1122 1115 1026 1040
901 1324 818 1250 1203 1078 890 1303 1147 1289
1187 1067 1118 1037 958 760 1101 949 883 699
824 643 980 935 878 934 910 1058 867 1083
844 814 1103 1000 788 1143 935 1069 990 880
1037 1151 863 990 1035 1112 931 970 1258 1029

7. The number of goals scored in 57 English and Scottish league matches for Saturday 23rd September, 1969, was:

1 1 0 2 3 3 5 2 2 1 4 4
2 3 3 3 2 2 1 0 4 6 2 5
6 1 4 1 4 3 4 2 7 6 1 2
3 6 4 2 1 4 3 3 3 6 8 3
5 3 3 2 1 3 5 3 5

8. The intelligence quotients of 100 children are given below:†

75	112	100	116	99	111	85	82	108	85
94	91	118	103	102	133	98	106	92	102
115	109	100	57	108	77	94	121	100	107
104	67	111	88	87	97	102	98	101	88
90	93	85	107	80	106	120	91	101	103
109	100	127	107	112	98	83	98	89	106
79	117	85	94	119	93	100	90	102	87
95	117	142	94	93	72	98	105	122	104
104	79	102	104	107	97	100	109	103	107
106	96	83	107	102	110	102	76	98	88

9. The sales value for the last 30 periods of a non-seasonal product are given below in units of £100:

43	41	74	61	79	60	71	69	63	77
70	66	64	71	71	74	56	74	41	71
63	57	57	68	64	62	59	52	40	76

10. The records of the total score of three dice in 100 throws are given below:

16	4	9	12	11	8	15	13	12	13
8	7	6	13	10	11	16	14	7	12
14	14	4	13	9	12	8	10	12	14
8	4	10	6	9	10	13	12	13	13
16	7	13	12	9	8	10	11	12	10
15	12	4	16	10	9	13	10	9	12
9	4	14	13	7	6	11	9	15	8
5	12	7	6	7	13	13	11	13	14
12	7	10	12	12	12	13	9	16	4

2.4 Solutions to Problems

1. Range = 0.11 – 0.06 = 0.05 min.

Since only two significant figures are given in the data, there is no choice regarding the class interval width.

Size of class interval = 0.01 min, giving only six class intervals (below the preferred minimum of eight).

† These data were taken from *Facts from Figures* by M. J. Moroney, Pelican.

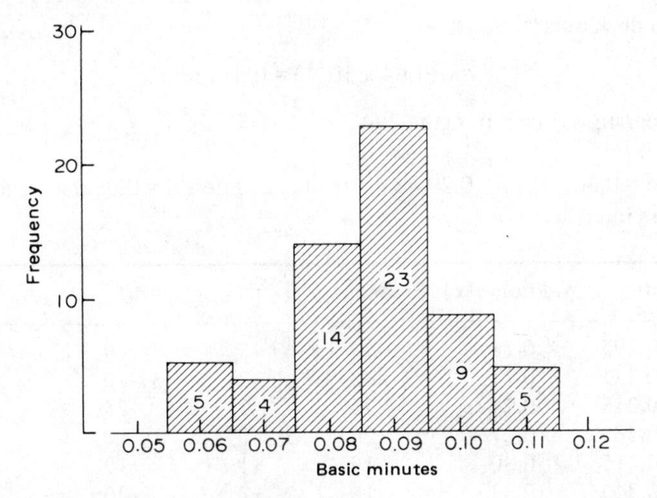

Figure 2.6. Glueing labels onto biscuit tins.

Class interval	Mid point (x)	Frequency (f)	u	uf	u^2f
0.055–0.065	0.06	5	−3	−15	45
0.065–0.075	0.07	4	−2	−8	16
0.075–0.085	0.08	14	−1	−14	14
0.085–0.095	0.09	23	0	0	0
0.095–0.105	0.10	9	+1	+9	9
0.105–0.115	0.11	5	+2	+10	20
		$\Sigma f = 60$		$\Sigma uf = -18$	$\Sigma u^2 f = 104$

Table 2.5

Transforming $x = x_0 + cu$

Let $x_0 = 0.09$
$c = 0.01$

1st moment about the origin = arithmetical mean,

$$\bar{x} = x_0 + c\frac{\Sigma uf}{\Sigma f} = 0.09 + 0.01 \times \left(\frac{-18}{60}\right) = 0.087 \text{ min}$$

Variance of

$$(s')^2 = c^2 \left[\frac{\Sigma u^2 f - \dfrac{(\Sigma uf)^2}{\Sigma f}}{\Sigma f}\right] = 0.01^2 \left[\frac{104 - \left(\dfrac{-18}{60}\right)^2}{60}\right] = 0.01^2 \left(\frac{104 - 5.4}{60}\right)$$

$$= 1.64 \times 10^{-4}$$

Standard deviation

$$s' = \sqrt{(1.64 \times 10^{-4})} = 0.013 \text{ min}$$

The histogram is shown in figure 2.6.

2. Range = 0.46 – 0.17 = 0.29 min; size of class interval = 0.03 min, giving 9–10 class intervals.

Class interval	Mid point (x)	Frequency (f)	u	uf	u^2f
0.165–0.195	0.18	3	−3	−9	27
0.195–0.225	0.21	5	−2	−10	20
0.225–0.255	0.24	25	−1	−25	25
0.255–0.285	0.27	26	0	0	0
0.285–0.315	0.30	19	+1	+19	19
0.315–0.345	0.33	10	+2	+20	40
0.345–0.375	0.36	3	+3	+9	27
0.375–0.405	0.39	0	+4	0	0
0.405–0.435	0.42	1	+5	+5	25
0.435–0.465	0.45	1	+6	+6	36
		$\Sigma f = 93$		$\Sigma uf = +15$	$\Sigma u^2 f = 219$

Table 2.6

(For histogram see figure 2.7.)

Calculation of the Mean and Standard Deviation

Transform

$$x = x_0 + cu$$

Let

$$x_0 = 0.27, \qquad c = 0.03$$

Average time

$$\bar{x} = x_0 + c\frac{\Sigma uf}{\Sigma f} = 0.27 + \left(0.03 \times \frac{15}{93}\right) = 0.275$$

Variance of sample

$$(s')^2 = c^2 \left[\frac{\Sigma u^2 f - \dfrac{(\Sigma uf)^2}{\Sigma f}}{\Sigma f}\right] = 0.03^2 \left[\frac{219 - \dfrac{(+15)^2}{93}}{93}\right] = 0.03^2 \left(\frac{219 - 2.42}{93}\right)$$

$$= \frac{0.03^2 \times 216.58}{93} = 0.0021$$

∴ Standard deviation $s' = 0.046$ min

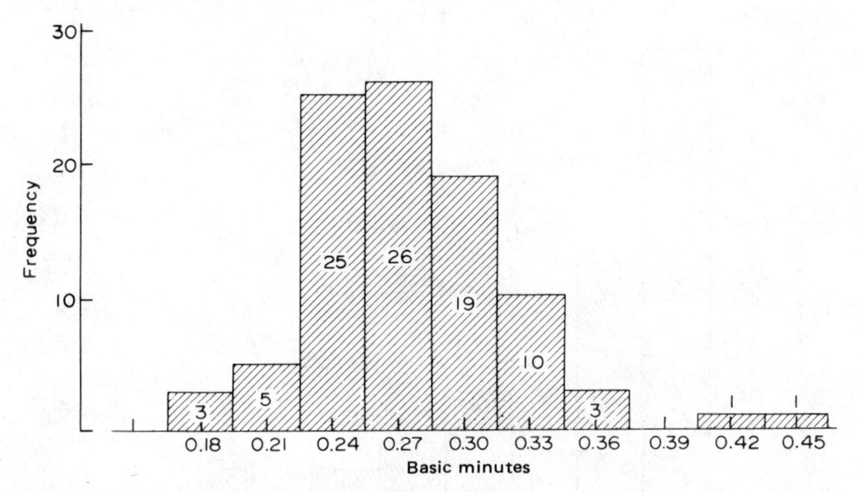

Figure 2.7. Time taken to assemble Hoover agitator.

3. Range = 4.60 – 0.01 = 4.59 min; width of class interval = 0.5 min.

Class interval	Frequency (f)	u	uf	u^2f
0–0.499	19	−2	−38	76
0.50–0.999	11	−1	−11	11
1.00–1.499	7	0	0	0
1.50–1.999	6	+1	+6	6
2.00–2.499	4	+2	+8	16
2.50–2.999	3	+3	+9	27
3.00–3.499	2	+4	+8	32
3.50–3.999	3	+5	+15	75
4.00–4.499	0	+6	+0	0
4.50–4.999	1	+7	+7	49
	56		$\Sigma uf = +4$	$\Sigma u^2 f = 292$

Table 2.7

(For histogram, see figure 2.8.)

Transform

$$x = x_0 + cu$$

Let

$$x_0 = 1.25, \qquad c = 0.50$$

1st moment about the origin = arithmetic average,

$$\bar{x} = x_0 + c\frac{\Sigma uf}{\Sigma f} = 1.25 + 0.50 \times \frac{4}{56} = 1.29 \text{ min}$$

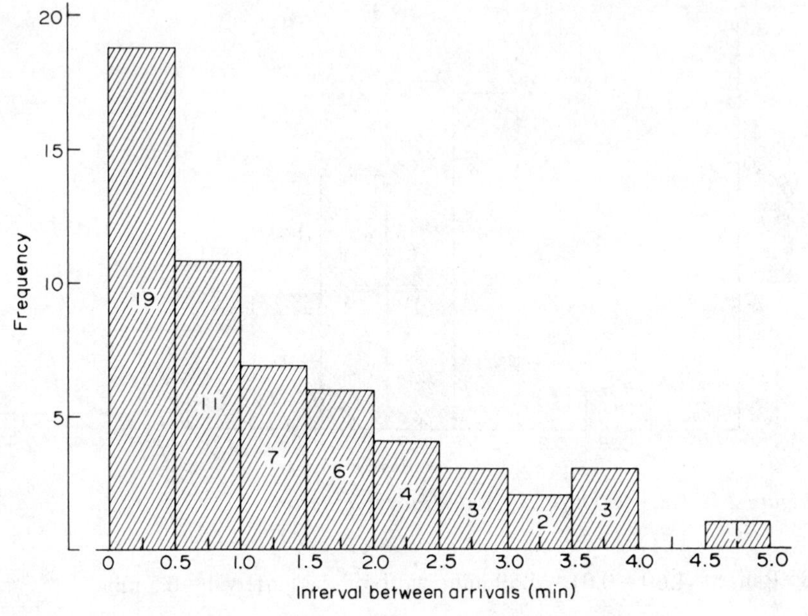

Figure 2.8. Interval between arrival of customers.

Variance of the sample

$$(s')^2 = c^2 \left[\frac{\Sigma u^2 f - \dfrac{(\Sigma uf)^2}{\Sigma f}}{\Sigma f}\right] = 0.5^2 \left[\frac{292 - \dfrac{(+4)^2}{56}}{56}\right] = 0.25\left(\frac{292 - 0.29}{56}\right) = 1.30$$

Standard deviation of sample $s' = 1.14$ min

4. Range = 9 – 0 = 9 defectives; width of class interval = 1 defective.

Number of defectives	Number of shifts (f)	u	uf	u^2f
0	3	–3	–9	27
1	7	–2	–14	28
2	9	–1	–9	9
3	12	0	0	0
4	9	+1	+9	9
5	6	+2	+12	24
6	3	+3	+9	27
7	2	+4	+8	32
8	0	+5	+0	0
9	1	+6	+6	36
	$\Sigma f = 52$		$\Sigma uf = +12$	$\Sigma u^2 f = 192$

Table 2.8

(For histogram see figure 2.9.)

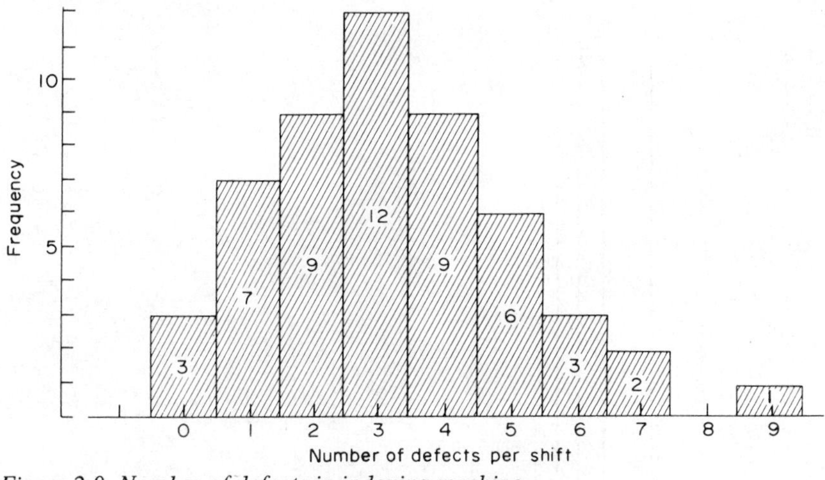

Figure 2.9. Number of defects in indexing machine.

Transform $x_0 = 3$

Let $c = 1$ defective

$\therefore$ Average number of defectives per shift $= 3 + 1 \times \dfrac{12}{52} = 3.2$ per shift

Variance of the sample

$$(s')^2 = 1^2 \times \left[\frac{192 - \dfrac{(12)^2}{52}}{52}\right] = 3.64$$

$\therefore$ Standard deviation = 1.9

5. Range = 192 – 4 = 188 min.

In this case, if equal class interval widths were chosen, then a width of perhaps 20 min would be suitable. However, as can be checked, in the case of the J-shaped distribution unequal class intervals give a better summary.

Class interval	Mid point (x)	Frequency (f)	u	uf	u^2f
0– 9.99	5	35	–3	–105	315
10– 19.99	15	30	–2	–60	120
20– 29.99	25	15	–1	–15	15
30– 39.99	35	6	0	0	0
40– 49.99	45	3	+1	3	3
50– 69.99	60	4	+2.5	10	25
70– 99.99	85	2	+5	10	50
100–139.99	120	3	+8.5	25.5	216.75
140–199.99	140	2	+13.5	27	364.5
		$\Sigma f = 100$		$\Sigma uf = -104.5$	$\Sigma u^2 f = 1108.5$

Table 2.9

(For histogram, see figure 2.10.)

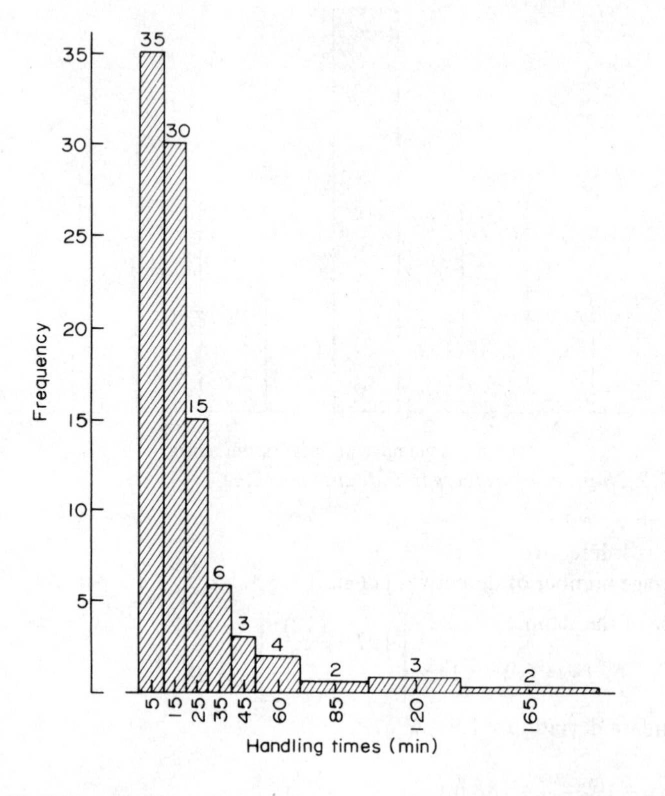

Figure 2.10. Crane handling times.

Transform $x = x_0 + cu$

Let $c = 10$ min

$x_0 = 35$

$$\text{Arithmetic average} = 35 - 10 \times \frac{104.5}{100} = 24.6 \text{ min}$$

Variance of the sample

$$(s')^2 = 10^2 \left[\frac{1082 - \dfrac{(-104.5)^2}{100}}{100} \right] = 10^2(9.72) = 972$$

Standard deviation $s' = 31.2$ min

6. Range = 1333 – 643 = 690 h; class interval chosen as 100.

Variate (x)	Frequency (f)	u	uf	u^2f
549.5– 649.5	1	−4	−4	16
649.5– 749.5	1	−3	−3	9
749.5– 849.5	10	−2	−20	40
849.5– 949.5	26	−1	−26	26
949.5–1049.5	26	0	0	0
1049.5–1149.5	18	+1	+18	18
1149.5–1249.5	11	+2	+22	44
1249.5–1349.5	7	+3	+21	63
	$\Sigma f = 100$		$\Sigma uf = +8$	$\Sigma u^2 f = 216$

Table 2.10

(For histogram, see figure 2.11.)

Transforming $x = x_0 + uc$

where $c = 100$ h
$x_0 = 1000$ h

Average life of bulbs,

$$\bar{x} = 1000 + 100 \times \left(\frac{+8}{100}\right) = 1008 \text{ h}$$

Variance of the sample

$$(s')^2 = 100^2 \left[\frac{216 - \dfrac{(+8)^2}{100}}{100}\right] = 21\,536$$

$\therefore$ Standard deviation $s' = 146.6$ h

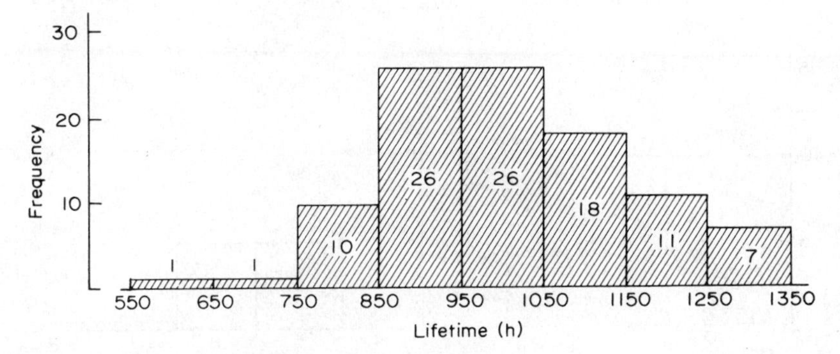

Figure 2.11. Lifetime of electric light bulbs.

7. Range = 0-8 goals.

Discrete distribution

∴ Width of class interval = 1 goal

Number of goals/match	Frequency (f)	u	uf	u^2f
0	2	−4	−8	32
1	9	−3	−27	81
2	11	−2	−22	44
3	15	−1	−15	15
4	8	0	0	0
5	5	+1	+5	5
6	5	+2	+10	20
7	1	+3	+3	9
8	1	+4	+4	16
	$\Sigma f = 57$		$\Sigma uf = -50$	$\Sigma u^2 f = 222$

Table 2.11

(For histogram, see figure 2.12.)

$$x_0 = 4 \qquad c = 1$$

∴ Average goals/match

$$\bar{x} = 4 + 1 \times \left(\frac{-50}{57}\right) = 3.12$$

Variance of sample

$$(s')^2 = 1^2 \times \left[\frac{222 - \frac{(50)^2}{57}}{57}\right] = 3.13$$

∴ Standard deviation of sample = 1.8

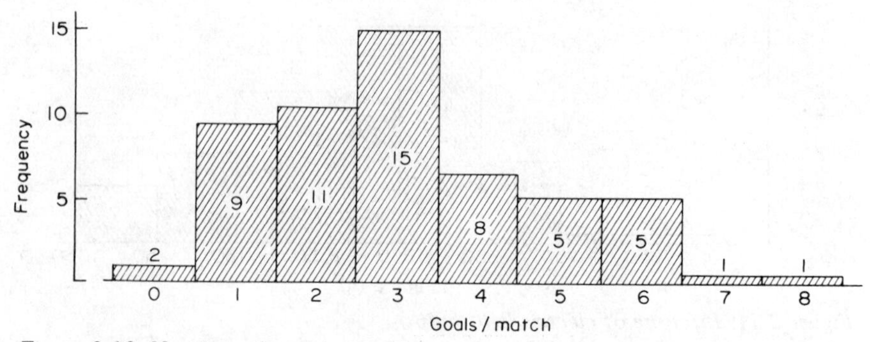

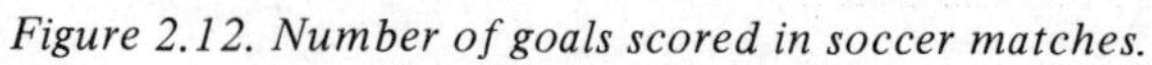

Figure 2.12. Number of goals scored in soccer matches.

8. Range = 143 – 57 = 85.

Suitable class intervals could be either 10 or 15. In this case as with the author in *Facts from Figures* the class interval is chosen as 10.

Class interval	Frequency (f)	u	uf	u^2f
54.5– 64.5	1	–4	–4	16
64.5– 74.5	2	–3	–6	18
74.5– 84.5	9	–2	–18	36
84.5– 94.5	22	–1	–22	22
94.5–104.5	33	0	0	0
104.5–114.5	22	+1	+22	22
114.5–124.5	8	+2	+16	32
124.5–134.5	2	+3	+6	18
134.5–144.5	1	+4	+4	16
	$\Sigma f = 100$		$\Sigma uf = -2$	$\Sigma u^2 f = 180$

Table 2.12

Transforming $x = x_0 + cu$ (For histogram, see figure 2.13.)

where $x_0 = 99.5$

$c = 10$

∴ Average intelligence quotient

$$\bar{x} = 99.5 + 10 \times \left(\frac{-2}{100}\right) = 99.3$$

Variance of sample

$$(s')^2 = 10^2 \left[\frac{180 - \frac{(-2)^2}{100}}{100}\right] = 180.0$$

∴ Standard deviation of sample $s' = 13.4$

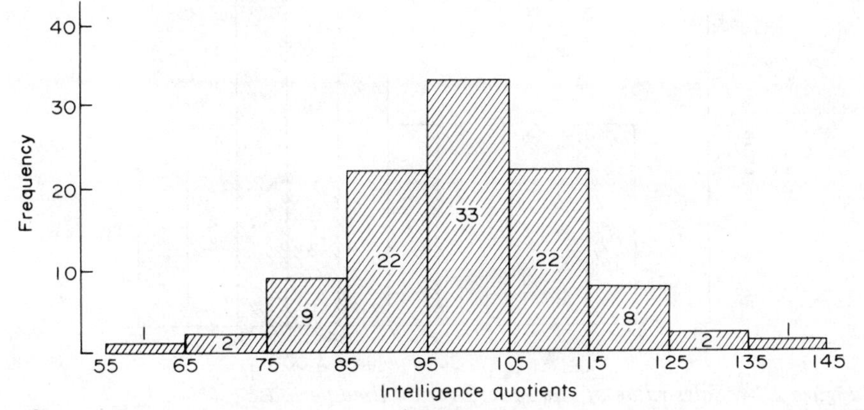

Figure 2.13. Intelligence quotients of children.

9. Range = 79 – 40 = 39; class interval width = 4.

Variate (x)	Frequency (f)	u	uf	u^2f
39.5–43.5	4	–5	–20	100
43.5–47.5	0	–4	0	0
47.5–51.5	0	–3	0	0
51.5–55.5	1	–2	–2	4
55.5–59.5	4	–1	–4	4
59.5–63.5	5	0	0	0
63.5–67.5	3	+1	+3	3
67.5–71.5	7	+2	+14	28
71.5–75.5	4	+3	+12	36
75.5–79.5	2	+4	+8	32
	$\Sigma f = 30$		$\Sigma uf = +11$	$\Sigma u^2 f = 207$

Table 2.13

(For histogram, see figure 2.14.)

where $x_0 = 61.5$

$c = 4$

Average sales/period

$$\bar{x} = 61.5 + 4\left(\frac{11}{30}\right) = 63$$

Variance of sample

$$(s')^2 = 4^2 \left[\frac{207 - \frac{(+11)^2}{30}}{30}\right] = 108.3$$

∴ Standard deviation of sample $s' = 10.4$

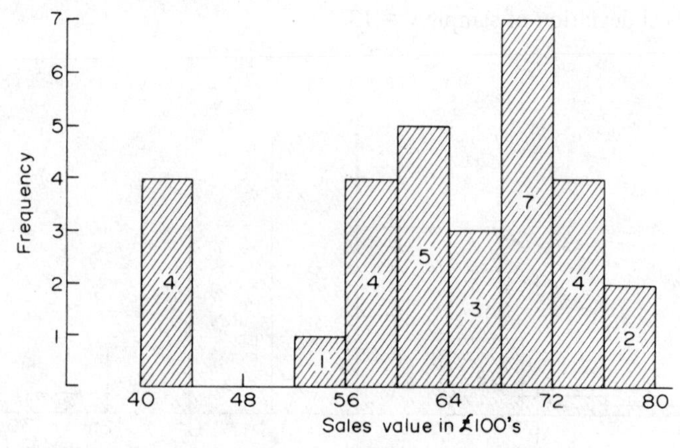

Figure 2.14. Sales value of a product over 30 time periods.

10. Range = 16 − 4 = 12; use class interval of 2 units.

Variate (x)	Frequency (f)	u	uf	u^2f
3.5– 5.5	7	−3	−21	63
5.5– 7.5	13	−2	−26	52
7.5– 9.5	17	−1	−17	17
9.5–11.5	18	0	0	0
11.5–13.5	29	+1	+29	29
13.5–15.5	11	+2	+22	44
15.5–17.5	5	+3	+15	45
	$\Sigma f = 100$		$\Sigma uf = +2$	$\Sigma u^2 f = 250$

Table 2.14

(For histogram, see figure 2.15.)

where $x_0 = 10.5$
$c = 2$

Average score

$$\bar{x} = 10.5 + 2\left(\frac{2}{100}\right) = 10.54$$

Variance of scores

$$(s')^2 = 2^2\left[\frac{250 - \left(\frac{2}{100}\right)^2}{100}\right] = 10$$

$\therefore$ Standard deviation $s' = 3.16$

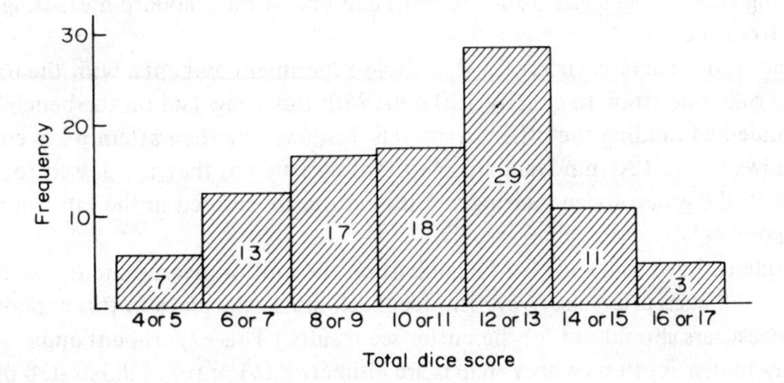

Figure 2.15. Total score of three dice.

2.5 Practical Laboratory Experiments and Demonstrations

The three experiments described are experiments 4, 5 and 6 from the authors' *Laboratory Manual in Basic Statistics*, pages 20–32.

As explained on page 20 of the manual, the authors leave the selection of the populations to be used to the individual instructor—use whatever is most suitable.

The objects of these experiments are firstly to show basic concepts involved and secondly to give experience in computing means and standard deviation. Thus data collection should be as quick as possible and the following points noted:

(1) How accurately should students measure? Obviously the unit of measurement must be small enough to give approximately eight to twenty class intervals and since the sample size of 50 is relatively small, the best number of class intervals is at the bottom end of the range.

(2) Selection of class intervals: The tables for computing mean and standard deviations are set out fully in the manual. However, with the kit, one obvious and quick experiment is to measure 50 rods from either the red or yellow population using the measuring rules. Again one of the experiments designed by the authors is described below. This 'straw' experiment is perhaps one of the best and most famous distribution experiments due to various points which can be made. Also described is the shove halfpenny experiment.

2.5.1 *The Drinking Straw Experiment*

The simplicity and speed of this experiment illustrate the main requirements of good design.

Here students (in groups of two or three) are given 50 or more ordinary drinking straws (usual size 180–250 mm) and one of the standard measuring rules from the kit.

One student acts as cutter for the whole experiment and cuts, with the use of the rule, one straw to exactly 130 mm. With this straw laid on the bench as the guide and holding the other straws 0.5–1 m away he then attempts to cut 50 straws to the 130 mm standard. As the straws are cut they are passed to others in the group for measuring and the results are entered in the table in the manual.

Students have to decide (or be guided) on the unit of measurement, i.e. at least 6 to 8 class intervals. (*Note:* no feedback must take place in this experiment and measurers should not let the cutter see results.) This experiment ends usually in distributions whose shapes are either (*a*), (*b*) or (*c*) as illustrated in figure 2.16.

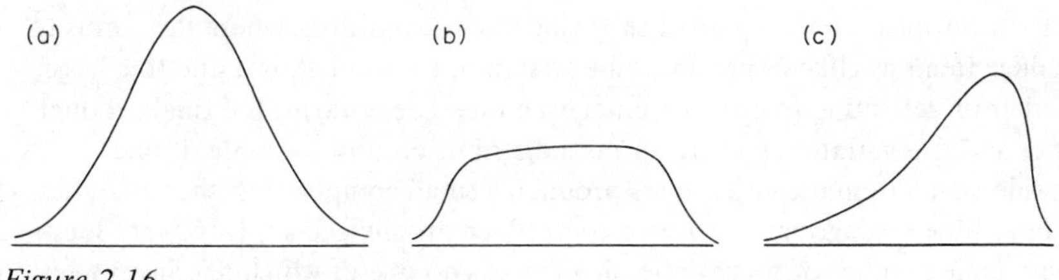

Figure 2.16

For the case of

(*a*) The cutter has held the standard and produced a bell-shaped curve.
(*b*) Here either consciously or not, the standard has been changing.
(*c*) Here the negative skew distribution has arisen by the cutter again either consciously or not, placing control on the short end of the straw.

2.5.2 *The Shove Halfpenny Experiment*

Number of persons: groups of 2 or 3.

Laboratory Equipment

Shove-halfpenny board or specially designed board (available from Technical Prototypes (Sales) Ltd).

Method

After one trial, carry out 50 further trials, measuring the distance travelled each time, the object being to send the disc the same distance at each trial.

Analysis

Summarise the data into a distribution, draw a histogram and calculate the mean and standard deviation.

2.5.3 *The Quincunx*

This use of a quincunx developed by the authors is outlined below and gives an effective simple demonstration of the basic concepts of variation.

This simple model, the principle of which was originally devised by Galton to give a mechanical representation of the binomial distribution, is an effective means of demonstrating to students the basic concepts of distributions.

The quincunx supplied with the statistical kit has ten rows of pins and seed is fed in a stream through the pattern of pins. The setting is such that each pin has a 50% chance of throwing any one seed to the right or to the left and thus, a symmetrical distribution is obtained. With this large array of pins and the speed of the stream, a simple analogy of the basic concept of distributions can be demonstrated speedily and effectively.

Distributions can be regarded as arising under conditions where the aim is to produce items as alike as possible (the stream in the model), but due to a large number of deflecting forces (the pins) each one independent, the final product varies and this variation pattern forms a distribution. For example, if one considers an automatic lathe, mass producing small components, then material and machine settings are controlled to produce products alike. However, due to a very large number of small deflecting forces–no one of which has an appreciable effect, otherwise it would be possible to correct for it–such as vibration, particles of dirt in the cooling oil, small random variation in the material, the final components give rise to a distribution similar to that generated by the quincunx.

3 Hypergeometric, Binomial and Poisson distributions

3.1 Syllabus Covered

Use of hypergeometric distribution; binomial distribution and its application; Poisson distribution and its application; fitting of distributions to given data.

3.2 Résumé of Theory and Concepts

Here only a brief résumé of theory is given since this is already easily available in a wide range of textbooks. However, this résumé, gives not only the basic laws but also the basic concepts, which are necessary to give a fuller understanding of the laws.

3.2.1 *The Hypergeometric Law*

If a group contains N items of which M are of one type and the remainder $N-M$, are of another type, then the probability of getting exactly x of the first type in a random sample of size n is

$$P(x) = \frac{\binom{M}{x}\binom{N-M}{n-x}}{\binom{N}{n}}$$

3.2.2 *The Binomial Law*

If the probability of success of an event in a single trial is p and p is constant for all trials, then the probability of x successes in n independent trials is

$$P(x) = \binom{n}{x} p^x (1-p)^{n-x}$$

3.2.3 *The Poisson Law*

If the chance of an event occurring at any instant is constant in a continuum of time, and if the average number of successes in time t is m, then the probability of x successes in time t is

$$P(x) = \frac{m^x e^{-m}}{x!}$$

where m = expected (or average) number of successes
e = exponential base e $\simeq$ 2.178

Here, the event's happening has a definite meaning but no meaning can be attached to its *not happening.* For example, the number of times lightning strikes can be counted and have a meaning but this is not true of the number of times lightning does *not* strike.

Again the Poisson law can be derived as the limit of the binomial under conditions where $p \to 0$ and $n \to \infty$ but such that np remains finite and equal to m.

Then the probability of x successes is

$$P(x) = \frac{m^x e^{-m}}{x!}$$

as before.

3.2.4 The essential requirement is for students to be able to decide which distribution is applicable to the problem, and for tnis reason the problems are all given under the general headings, because for example the statement that the problems relate to the Poisson law almost defeats the purpose of setting the problems.

Tutors must stress the relationship between these distributions so that students can understand the type to use for any given situation.

Tutors can introduce students to the use of binomial distribution in place of hypergeometric distribution in sampling theory when $n/N \leqslant 0.10$.

Students should be introduced to the use of statistical tables at this stage. For all examples and problems, the complementary set of tables, namely *Statistical Tables* by Murdoch and Barnes, published by Macmillan, has been used. As mentioned in the preface, references to these tables will be followed by an asterisk.

Note: The first and second moments of the binomial and Poisson distributions are given below.

	Binomial	*Poisson*
1st moment (mean) μ	np	m
2nd moment about the mean (variance) σ^2	$np(1-p)$	m

3.2.5 *Examples on Use of Distributions*

1. Assuming randomness in shuffling, what is the distribution of the number of diamonds in a 13-card hand? What is the probability of getting exactly five diamonds in the hand?

This is the hypergeometric distribution.

Type 1—diamonds 13 cards
Type 2—not diamonds 39 cards

Probability of x diamonds in 13-card hand is

$$P(x) = \frac{\binom{13}{x}\binom{39}{13-x}}{\binom{52}{13}}$$

Probability of exactly five diamonds in the hand

$$P(5) = \frac{\binom{13}{5}\binom{39}{8}}{\binom{52}{13}} = \frac{\frac{13!}{8!5!} \times \frac{39!}{31!8!}}{\frac{52!}{39!13!}} = 0.1247$$

2. A distribution firm has 50 lorries in service delivering its goods; given that lorries break down randomly and that each lorry utilisation is 90%, what proportion of the time will

(*a*) exactly three lorries be broken down?
(*b*) more than five lorries be broken down?
(*c*) less than three lorries be broken down?

This is the binomial distribution since the probability of success, i.e. the probability of a lorry breaking down, is $p = 0.10$ and this probability is constant. The number of trials $n = 50$.

(*a*) Probability of exactly three lorries being broken down

$$P(3) = \binom{50}{3} 0.10^3 \, (1-0.10)^{47}$$

from table 1* of statistical tables

$$\therefore \quad P(3) = 0.8883 - 0.7497 = 0.1386$$

(*b*) Probability of more than five lorries being broken down

$$P(x > 5) = \sum_{x=6}^{\infty} \binom{50}{x} 0.10^x (1-0.10)^{50-x}$$

from table 1*

$$P(x > 5) = 0.3839$$

(*c*) Probability of less than three lorries being broken down

$$P(x < 3) = 1 - \sum_{x=3}^{\infty} \binom{50}{x} 0.10^x (1-0.10)^{50-x} = 1 - 0.8883 = 0.1117$$

3. How many times must a die be rolled in order that the probability of 5 occurring is at least 0.75?

This can be solved using the binomial distribution. Probability of success, i.e. a 5 occurring, is $p = \frac{1}{6}$.

Let k be the unknown number of rolls required, then probability of x number of 5's in k rolls

$$P(x) = \binom{k}{x}\left(\frac{1}{6}\right)^x\left(\frac{5}{6}\right)^{k-x}$$

$\therefore$ Probability required is

$$\sum_{x=1}^{k} \binom{k}{x}\left(\frac{1}{6}\right)^x\left(\frac{5}{6}\right)^{k-x} = 0.75$$

Since

$$1 - \sum_{x=1}^{k} \binom{k}{x}\left(\frac{1}{6}\right)^x\left(\frac{5}{6}\right)^{k-x} = \text{probability of not getting a 5 in } k \text{ throws} = \left(\frac{5}{6}\right)^k$$

$$\therefore \quad \frac{5}{6}^{\,k} = 1 - 0.75 = 0.25$$

$$\therefore \quad k = \frac{\log 0.25}{\log 0.833} = \frac{\bar{1}.39794}{\bar{1}.92065} = \frac{0.60206}{0.07935} = 7.6$$

$\therefore$ Number of throws required = 8

4. A firm receives very large consignments of nuts from its supplier. A random sample of 20 is taken from each consignment. If the consignment is in fact 30% defective, what is

(*a*) probability of finding no defective nuts in the sample?
(*b*) probability of finding five or more defective nuts in the sample?

This is strictly a hypergeometric problem but it can be solved by using the binomial distribution since probability of success, i.e. of obtaining a defect, is $p = 0.30$ which can be assumed constant. The consignment is large enough to ignore the very slight change in p as the sample is taken.

(*a*) From table 1*, $p = 0.30, n = 20$

$$P(0) = \binom{20}{0} 0.30^{\circ} (1-0.30)^{20} = 1-0.9992 = 0.0008$$

(*b*) The probability of finding five or more defectives

$$P(x \geqslant 5) = \sum_{x=5}^{20} \binom{20}{x} 0.30^{x} (1-0.30)^{20-x}$$

from table 1* = 0.7625

5. The average usage of a spare part is one per month. Assuming that all machines using the part are independent and that breakdowns occur at random, what is

(*a*) the probability of using three spares in any month?
(*b*) the level of spares which must be carried at the beginning of each month so that the probability of running out of stock in any month is at most 1 in 100?

This is the Poisson distribution.

The expected usage $m = 1.0$

(*a*) ∴ Probability of using three spares in any month

$$P(3) = \frac{1.0^{3} e^{-1.0}}{3!}$$

from table 2* $P(3) = 0.0803 - 0.0190 = 0.0613$

(*b*) This question is equivalent to: what demand in a month has a probability of at most 0.01 of being equalled or exceeded?

Note: Runout if stock falls to zero.

From table 2*

Stocking four spares, probability of four or more = 0.0190
Stocking five spares, probability of five or more = 0.0037
∴ Stock five, the probability of runout being 0.0037

(*Note:* It is usual to go to a probability of 1/100 or less.)

6. The average number of breakdowns due to failure of a bearing on a large automatic indexing machine is two per six months. Assume the failures are

random, and calculate and draw the probability distribution of the number of failures per six months per machine over 100 machines.

Calculate the average and the standard deviation of the distribution.

This is the Poisson distribution.

Expected number of failures per machine per six months, $m = 2$.

Number of failures	Probability P_i	Expected number of failures f_i	u	uf_i	$u^2 f_i$
0	0.1353	13.5	−2	−27	54.0
1	0.2707	27.1	−1	−27.1	27.1
2	0.2707	27.1	0	0	0
3	0.1804	18.0	+1	+18	18.0
4	0.0902	9.0	+2	+18	36.0
5	0.0361	3.6	+3	+10.8	32.4
6	0.0121	1.2	+4	4.8	19.2
7 or over	0.0045	0.5	+5	2.5	12.5
	1.0000	$\Sigma f_i = 100$		$\Sigma uf_i = 0$	$u^2 f_i = 199.2$

Table 3.1. The values have been calculated from table 2* of statistical tables.

Transform $x = uc + x_0$

$x_0 = 2$

$c = 1$

The arithmetical average

$$\bar{x} = x_0 + c\frac{\Sigma uf}{\Sigma f} = 2$$

Variance

$$(s')^2 = c^2\left[\frac{\Sigma u^2 f - \frac{(\Sigma uf)^2}{\Sigma f}}{\Sigma f}\right] = 1^2 \times \left(\frac{199.2 - 0}{100}\right) = 1.992$$

Standard deviation = 1.41

3.2.6 *Special Examples of the Poisson Distribution of General Interest*

The following examples have been chosen to show the use of the Poisson distribution and to illustrate clearly the tremendous potential of statistics or, that is, the logic of inference.

Students will be introduced here to some of the logic used later so that they can see, even at this introductory stage, something of the overall analysis using statistical methods.

1. Goals Scored in Soccer

Problem 7 in chapter 2 (page 47) gives the data to illustrate this example. The distribution of actual goals scored in the 57 matches is given in table 3.2. The mean of this distribution is easily calculated, as in chapter 2, as

average number of goals/match m = 3.1

Number of goals/match	0	1	2	3	4	5	6	7	8	Total
Frequency	2	9	11	15	8	5	5	1	1	57

Table 3.2

Setting up the hypothesis that goals occur at random at a constant average rate, i.e. it does not matter which team is playing, then the Poisson distribution should fit these data. Using table 2* of statistical tables the probabilities are given in table 3.3, together with the Poisson frequencies.

Number of goals/match	0	1	2	3	4	5	6	7	8	Total
Poisson probability distribution	0.045	0.140	0.217	0.223	0.173	0.107	0.056	0.024	0.0142	1.000
Poisson frequency distribution	2.6	7.98	12.4	12.7	9.9	6.1	3.2	1.4	0.8	57
Actual frequency distribution	2	9	11	15	8	5	5	1	1	57

Table 3.3

The agreement will be seen to be fairly close and when tested (see chapter 8), is a good fit. It is interesting to see that the greater part of the variation is due to this basic law of variation. However, larger samples tend to show that the Poisson does not give a correct fit in this particular context.

2. Deaths due to Horsekicks

The following example due to Von Bortkewitz gives the records for 10 army corps over 20 years, or 200 readings of the number of deaths of cavalrymen due to horsekicks. The frequency distribution of number of deaths per corps per year is shown in table 3.4.

Number of deaths/corps/year	0	1	2	3	4
Frequency	109	65	22	3	1

Table 3.4

From this table the average number of deaths/corps/year, $m = 0.61$

Setting up the null hypothesis, namely, that the probability of a death has been constant over the years and is the same for each corps, is equivalent to postulating that this pattern of variation follows the Poisson law. Fitting a Poisson distribution to these data and comparing the fit, gives a method of testing this hypothesis. Using table 2* of statistical tables, and without interpolating, i.e. use $m = 0.60$, gives the results shown in table 3.5

Number of deaths	0	1	2	3	4 or more	Total
Poisson probability	0.5488	0.3293	0.0988	0.0197	0.0034	1.0000
Poisson frequency	109.8	65.9	19.8	3.9	0.68	200
Actual frequency	109	65	22	3	1	200

Table 3.5

Comparison of the actual pattern of variation shows how closely it follows the basic Poisson law, indicating that the observed differences between the corps are entirely due to chance or a basic law of nature.

3. *Outbreaks of War*

The data in table 3.6 (from *Mathematical Statistics* by J. F. Ractliffe, O.U.P.) give the number of outbreaks of war each year between the years 1500 and 1931 inclusive.

Number of outbreaks of war	0	1	2	3	4	5	Total
Frequency	223	142	48	15	4	0	432

Table 3.6

Setting up a hypothesis that war was equally likely to break out at any instant of time during this 432-year period would give rise to a Poisson distribution. The fitting of this Poisson distribution to the data gives a method of testing this hypothesis.

The average number of outbreaks/year = $0.69 \doteqdot 0.70$

Using table 2* of statistical tables, table 3.7 gives a comparison of the actual variation with that of the Poisson. Again comparison shows the staggering fact that life has closely followed this basic law of variation.

Number of outbreaks of war	0	1	2	3	4	5 or more	Total
Poisson probability	0.4966	0.3476	0.1217	0.0283	0.0050	0.0008	1.00
Poisson frequency	214.5	150.2	52.6	12.2	2.0	0.3	432
Actual frequency	223	142	48	15	4	0	432

Table 3.7

4. Demand for Spare Parts for B-47 Aircraft Airframe

Item	Units demanded per week	Number of weeks	
		Observed	Poisson[a]
Seal: $3 each (1AFE 15-24548-501)	0	48	46
	1	12	16
	2	2	3[b]
	3	2	–
	50[c]	1	–
Mean demand per week 0.3[c]			
Dome assembly: $610 each (1AFE4-2608-826)	0	33	26
	1	17	24
	2	7	11
	3	5	3
	4	2	1[d]
	6	1	–
Mean demand per week 0.9			
Boost assembly–elevator control $800 each (1AFE 15-24377-27, and substitute -20, -504)	0	20	17
	1	22	23
	2	13	15
	3	5	7
	4	3	2
	5	2	1[e]
Mean demand per week 1.3			

Table 3.8. Observed frequencies of demand compared with derived Poisson distributions

[a] Computed by assuming that the observed mean demand per week is the mean of the Poisson distribution.

[b] Two units or more.

[c] A demand of 50 units by a single aircraft was recorded on 23 December 1953. The mean used to fit the Poisson distribution (0.3) was obtained omitting this demand.

[d] Four units or more.

[e] Five units or more.

The actual demands at the MacDill Airforce Base per week for three spares for B-47 airframe over a period of 65 weeks are given in table 3.8.

The Poisson frequencies are obtained by using the statistical tables and table 3.8 gives a comparison of the actual usage distribution with that of the Poisson distribution.

The theoretical elements assuming the Poisson distribution are shown in the table also. It will be seen that these distributions agree fairly well with actual demands.

5. Spontaneous Ignitions in an Explosives Factory

The distribution of the number of spontaneous ignitions per day in an explosives factory is shown in table 3.9 and covers a period of 250 days. The Poisson frequencies, using the same mean number of explosions per day, have been calculated and the fit found to be good. This implies that the explosions occur at random, thus making it very unlikely that there is any systematic cause of ignition.

Number of ignitions	Observed number of days	Poisson number of days
0	75	74.2
1	90	90.1
2	54	54.8
3	22	22.2
4	6	6.8
5	2	1.6
6 or more	1	0.4

Table 3.9. Mean number of ignitions per day = 1.126.

Authors' Special Note

In all the foregoing examples, the (actual) observed distribution is compared with the (theoretical) expected distribution assuming the null hypothesis to be true. It should be stressed here that the degree of agreement between the observed and theoretical distributions can only be assessed by special tests, called significance tests. These tests will be carried out in chapter 8 later in the book.

3.3 Problems for Solution

1. A book of 600 pages contains 600 misprints distributed at random. What is the chance that a page contains at least two misprints?

2. If the chance that any one of ten telephone lines is busy at any instant is 0.2, what is the chance that five of the lines are busy?

3. A sampling inspection scheme is set up so that a sample of ten components is taken from each batch supplied and if one or more defectives is found the batch is rejected. If the suppliers' batches are defective

(*a*) 10% and (*b*) 20% what percentage of the batches will be rejected?

4. In a group of five machines, which run independently of each other, the chance of a breakdown on each machine is 0.20. What is the probability of breakdown of 0, 1, 2, 3, 4, 5 machines? What is the expected number of breakdowns?

5. In a quality control scheme, samples of five are taken from the production at regular intervals of time.

What number of defectives in the samples will be exceeded 1/20 times if the process average defective rate is (*a*) 10%, (*b*) 20%, (*c*) 30%?

6. In a process running at 20% defective, how often would you expect in a sample of 20 that the rejects would exceed four?

7. From a group of eight male operators and five female operators a committee of five is to be formed. What is the chance of

(*a*) all five being male?
(*b*) all five being female?
(*c*) how many ways can the committee be formed if there is exactly one female on it?

8. In 1000 readings of the results of trials for an event of small probability, the frequencies f_i and the numbers x_i of successes were:

x_i	0	1	2	3	4	5	6	7
f_i	305	365	210	80	28	9	2	1

Show that the expected number of successes is 1.2 and calculate the expected frequencies assuming Poisson distribution.

Calculate the variance of the distribution.

3.4 Solutions to the Problems

1. Assuming an average of one misprint per page, use of Poisson table 2* gives

$$P(2 \text{ or more misprints}) = 0.2642$$

2. $P(5 \text{ lines busy}) = \binom{10}{5} 0.2^5\, 0.8^5 = 0.0264$ from table 1* in statistical tables.

3. (*a*) Sample size $n = 10$
   Probability of defective $= p = 0.10$
   Reject on one or more defectives in sample of 10
   From table 1*
   ∴ Probability of finding one or more defectives in 10 = 0.6513
   ∴ Percentage of batches rejected = 65.13

   (*b*) Sample size $n = 10$
   Probability of a defective $p = 0.20$
   Reject on one or more defectives in sample of 10
   From table 1*
   ∴ Probability of finding one or more defective in 10 = 0.8926
   ∴ Percentage of batches rejected = 89.26

4. $\left.\begin{matrix} n = 5 \\ p = 0.20 \end{matrix}\right\}$ From statistical tables the probabilities of 0,1,2,3,4,5 machines breaking down have been calculated and are given in table 3.10.

Number of machines broken down	Probability of this number	
0	0.33	approximately
1	0.41	
2	0.20	
3	0.05	
4	0.01	
5	0	

Table 3.10

Expected number of breakdowns = $np = 5 \times 0.20 = 1$

5. (*a*) $n = 5$
   $p = 0.10$

   From table 1*
   ∴ Probability of exceeding 1 = 0.0815
   ∴ Probability of exceeding 2 = 0.0086
   1 in 20 times is a probability of 0.05
   ∴ Number of defectives exceeded 1 in 20 times is greater than 1 but less than 2.

(*b*) $n = 5$
$p = 0.20$

From table 1*
$\therefore$ Probability of more than 2 = 0.0579
$\therefore$ Number of defectives exceeded 1 in 20 times (approximately) is 2

(*c*) $n = 5$
$p = 0.30$

From table 1*
$\therefore$ Probability of more than 3 = 0.0318
$\therefore$ Number of defectives exceeded 1 in 20 times is nearly 3

6. $n = 20$
$p = 0.20$
From table 1*
$\therefore$ Probability of more than four rejects = 0.3704
$\therefore$ Four will be exceeded 37 times in 100

7. (*a*) $M = 8, N - M = 5, N = 13, n = 5$
$x = 5$

Probability

$$= \frac{\binom{8}{5}\binom{5}{0}}{\binom{13}{5}} = \frac{\frac{8!}{5!\,3!}}{\frac{13!}{5!8!}} = \frac{8!\,5!\,8!}{5!\,3!\,13!} = \frac{8 \times 7 \times 6 \times 5 \times 4}{13 \times 12 \times 11 \times 10 \times 9} = 0.044$$

(*b*) $M = 8, N = 13, N - M = 5, n = 5$
$x = 0$

Probability

$$= \frac{\binom{8}{0}\binom{5}{5}}{\binom{13}{5}} = \frac{1 \times 1}{\frac{13!}{8!\,5!}} = \frac{8!\,5!}{13!} = \frac{5 \times 4 \times 3 \times 2 \times 1}{13 \times 12 \times 11 \times 10 \times 9} = 0.00078$$

(*c*) Number of ways one female can be chosen from five

$$= \binom{5}{1} = 5$$

Number of ways four males can be chosen from eight

$$= \binom{8}{4}$$

∴ Total number of ways

$$= 5 \times \binom{8}{4} = 5 \times \frac{8!}{4!\,4!} = \frac{5 \times 8 \times 7 \times 6 \times 5}{4 \times 3 \times 2 \times 1} = 350$$

8.

10.

	x	f	u	uf	u^2f
Assumed mean	0	305	−1	−305	305
	1	365	0	0	0
	2	210	+1	210	210
	3	80	+2	160	320
	4	28	+3	84	252
	5	9	+4	36	144
	6	2	+5	10	50
	7	1	+6	6	36
		Σf 1000		Σuf 201	$\Sigma u^2 f$ 1317

Table 3.11

$$\text{Expected number} = \bar{x} = 1 + \frac{\Sigma uf}{f} = 1.201$$

$$\text{Variance} = \frac{\Sigma u^2 f - \dfrac{(\Sigma uf)^2}{\Sigma f}}{\Sigma f} = \frac{1317 - 40}{1000} = 1.277$$

3.5 Practical Laboratory Experiments and Demonstrations

The authors feel that of the three distributions the binomial lends itself best to demonstration by laboratory experiments. Attempts to demonstrate a true hypergeometric or Poisson distribution tend to be either very tedious and/or relatively expensive.

However, with the use of the binomial sampling boxes† the basic concepts and mechanics of the binomial can be speedily and effectively demonstrated.

The use of both 6-sided and/or the special decimal dice also give a simple method for carrying out binomial distribution experiments.

Appendix 1 contains the full details of the experiment, together with a sample set of results.

† Available in two sizes from Technical Prototypes Ltd. 1A, Westholme Street, Leicester.

Appendix 1—Experiment 7 and Sample Results

Binomial Distribution

Number of persons: 2 or 3.

Object

The experiment is designed to demonstrate the basic properties of the binomial law.

Method

Using the binomial sampling box, take 50 samples of size 10 from the population, recording in table 18, the number of coloured balls found in each sample.

(*Note:* Proportion of coloured (i.e. other than white) balls is 0.15.)

Analysis

1. Group the data of table 18 into the frequency distribution, using the top part of table 19.
2. Obtain the experimental probability distribution of the number of coloured balls found per sample and compare it with the theoretical probability distribution.
3. Combine the frequencies for all groups, using the lower part of table 19, and obtain the experimental probability distribution for these combined results. Again, compare the observed and theoretical probability distributions.
4. Enter, in table 20, the total frequencies obtained by combining individual groups' results. Calculate the mean and standard deviation of this distribution and compare them with the theoretical values given by np and $\sqrt{[np(1-p)]}$ respectively where, in the present case, $n = 10$ and $p = 0.15$.

Sample Results

1–10	3	2	1	3	4	3	2	0	0	1
11–20	2	3	2	3	3	1	2	1	1	0
21–30	0	0	1	1	2	1	3	1	1	3
31–40	4	4	1	1	4	1	2	3	2	2
41–50	1	1	1	1	2	1	0	0	1	2

Table 3.12 (Table 18 of the laboratory manual)

Summarise these data in table 19.

		Number of coloured balls in sample										Total frequency
		0	1	2	3	4	5	6	7	8	9	10	
'Tally-marks' Group No.___		卌 ||	卌 卌 卌 ||||	卌 卌 |	卌 ||||	||||							
Experimental frequency		7	19	11	9	4							50
Experimental probability		0·14	0·38	0·22	0·18	0·08							1·0
Theoretical probability		0·197	0·347	0·276	0·13	0·04	0·008	0·001					1·0
Group results	1	7	19	11	9	4							50
	2	12	20	8	8	1	1						50
	3	10	16	17	4	3							50
	4	8	16	17	5	2	2						50
	5	7	5	17	13	7		1					50
	6	5	21	11	10	2	1						50
	7	12	16	12	9	1							50
	8	13	17	11	7	2							50
Total frequency (all groups)		74	130	104	65	22	4	1					400
Experimental probability		0·185	0·325	0·260	0·163	0·055	0·010	0·0025					

Table 3.13 (Table 19 of the laboratory manual)

Frequency f	Number of coloured balls per sample x	fx	fx^2
74	0	0	0
130	1	130	130
104	2	208	416
65	3	195	585
22	4	88	352
4	5	20	100
1	6	6	36
	7		
	8		
	9		
	10		
Totals $\Sigma f = 400$		$\Sigma fx = 647$	$\Sigma fx^2 = 1619$

Table 3.14 (Table 20 of the laboratory manual)

For the distribution of number of coloured balls per sample of 10

$$\text{observed mean} = \frac{\Sigma fx}{\Sigma f} = \frac{647}{400} = 1.618$$

$$\text{Observed standard deviation} = \sqrt{\left[\frac{\Sigma fx^2 - \frac{(\Sigma fx)^2}{\Sigma f}}{\Sigma f}\right]} = \sqrt{\frac{1619 - \frac{(647)^2}{400}}{400}}$$

$$= 1.19$$

Theoretical mean = $np = 10 \times 0.15 = 1.5$

Theoretical standard deviation = $\sqrt{[np(1-p)]} = \sqrt{(10 \times 0.15 \times 0.85)}$

$$= \sqrt{1.275} = 1.13.$$

4 Normal distribution

4.1 Syllabus Covered

Equation of the normal curve; area under the normal curve; ordinates of the normal curve; standardised normal variate; use of tables of area; fitting of normal distribution to data; normal probability paper.

4.2 Résumé of Theory

4.2.1 *Introduction*

The normal, or gaussian, distribution occupies a central place in the theory of statistics. It is an adequate, and often very good, approximation to other distributions which occur; examples of this are given in chapters 5 and 6. Many of the advanced methods of statistics require the assumption that the basic variables being used are normally distributed; the purpose of this is usually to allow standard tests of significance to be applied to the results.

It often happens, however, that data summarised into a frequency distribution (see chapter 2) are more or less normally distributed; that is, some central value of the variable has the highest frequency of occurrence and the class frequencies diminish near enough symmetrically on either side of the central value. In such cases, it is very convenient to use the properties of the normal distribution to describe the population. This chapter deals with the main properties of the normal distribution.

4.2.2 *Equation of the Normal Curve*

Chapter 2 mentioned that, the greater the number of readings that are taken, the more the outline of the plotted histogram tends to a smooth curve. If the population is actually normal then this limiting shape of the histogram will be similar to that in figure 4.1.

The curve can be described in terms of an equation so that the height of the curve, y, can be expressed in terms of the value of the measured variable, $\bar{x}$.

This equation is

$$y = \frac{1}{\sigma\sqrt{(2\pi)}} e^{-\frac{(x-\mu)^2}{2\sigma^2}}$$

where μ is the mean of the variable x
σ is the standard deviation of x
e is the well-known mathematical constant (=2.718 approximately)
π is another well-known mathematical constant (=3.142 approximately)

This equation can be used to derive various properties of the normal distribution. A useful one is the relation between area under the curve and deviation from the mean, but before looking at this we need to refer to a *standardised variable.*

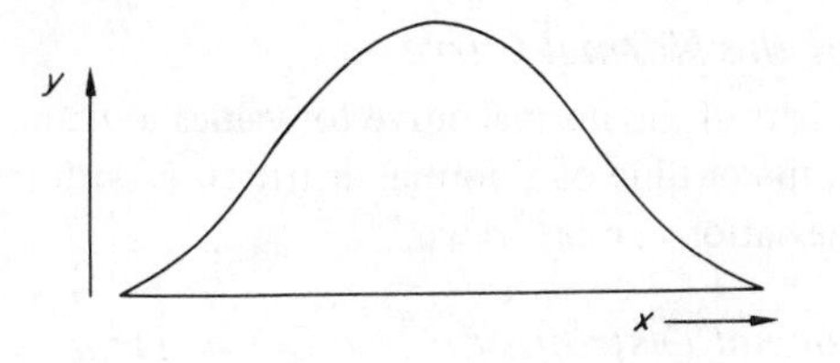

Figure 4.1

4.2.3 *Standardised Variate*

Any random variable, x, having mean, μ, and standard deviation, σ, can be expressed in standardised form, i.e. x is measured from μ in multiples of σ. The standardised variable is therefore given by $(x - \mu)/\sigma$ and is dimensionless.

In particular, if x is a *normal* variate then

$$u = \frac{x - \mu}{\sigma}$$

is a *standardised normal variate.*

Tables 3, 4 and 5* in statistical tables are tabulated in terms of this standardised normal variate, u, and therefore they apply to any normal variate.

4.2.4 *Area under Normal Curve*

The total area under the normal curve is unity (as is the case for any probability density function) and the area under the curve between two values of x, say a and b (shown shaded in figure 4.2) gives the proportion of the population having

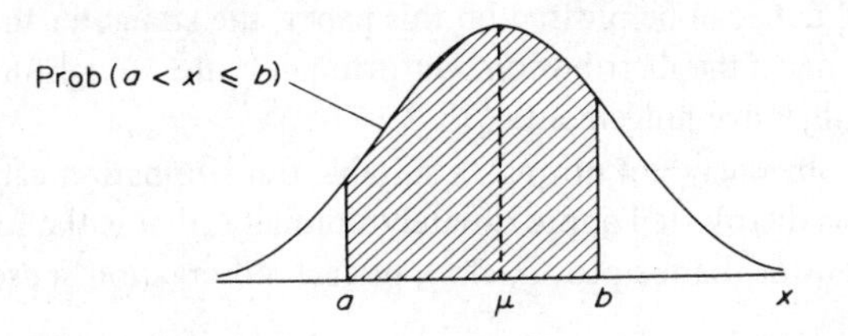

Figure 4.2

values between a and b. This is equal to the probability that a single random value of x will be bigger than a but less than b.

By standardising the variable and using the symmetry of the distribution, table 3* can be used to find this probability as well as the unshaded areas in each tail.

4.2.5 *Percentage Points of the Normal Distribution*

Table 4* gives percentage points (this is the common name although it is actually 'proportion points' which are tabulated) of the normal distribution; the α-proportion point or 100 α percentage point is the value of u, denoted by u_α, which is exceeded with probability α. Negative values of u_α corresponding to α greater than 0.50 can be found by symmetry.

4.2.6 *Ordinates of the Normal Curve*

Table 5* gives the height of the normal curve for values of u and by plotting a selection of points, the outline of a normal distribution with any required mean and standard deviation can be drawn.

4.2.7 *Fitting a Normal Distribution to a Set of Data*

Observed data will often be presented in the form of a frequency distribution together with a histogram. A normal distribution can be fitted to such a summary. The continuous curve outlining the shape of the normal distribution with the same (or any other) mean and standard deviation can be superimposed on the histogram using the ordinates in table 5*.

However, a more usual approach is to find the expected frequencies in each class interval of the observed data assuming that the population is normal with some given mean and standard deviation. This is best done using table 3* of areas and gives a basis for testing whether the assumption of normality is reasonable for the observed data (see chapter 8 for an example).

4.2.8 *Arithmetic Probability Paper*

This is graph paper with a special scale which makes the normal distribution, when plotted cumulatively, appear as a straight line. One axis has a linear scale and on this one convenient values of the variate are plotted. The other scale is usually marked in percentages which represent the probability that the variate takes on a value less than or equal to each of the plotted values.

Any observed data can be plotted on this paper, the straighter the line the more nearly normal is the distribution. Unfortunately the straightness of the line is rather a subjective judgement.

If a variate is obviously not normal, a suitable transformation can sometimes be found which is distributed approximately normally; that is the logarithm, say, (or the square root or the reciprocal, etc.), of each observation is used as the variable. By plotting these new variables on probability paper it can be seen whether any of the transformations gives a straight line.

4.2.9 *Worked Examples*

1. What is the chance that a random standardised normal variate

   (*a*) will exceed 1.0?
   (*b*) will be less than 2.0?
   (*c*) will be less than −2.0?
   (*d*) will be between −1.5 and +0.5?

Table 3* can be used to find these probabilities and it is useful to draw a diagram to ensure that the appropriate areas are found. In figure 4.3 the shaded areas represent the required answer. Remember that table 3* gives the probability of *exceeding* the specified value of u for positive values of u only.

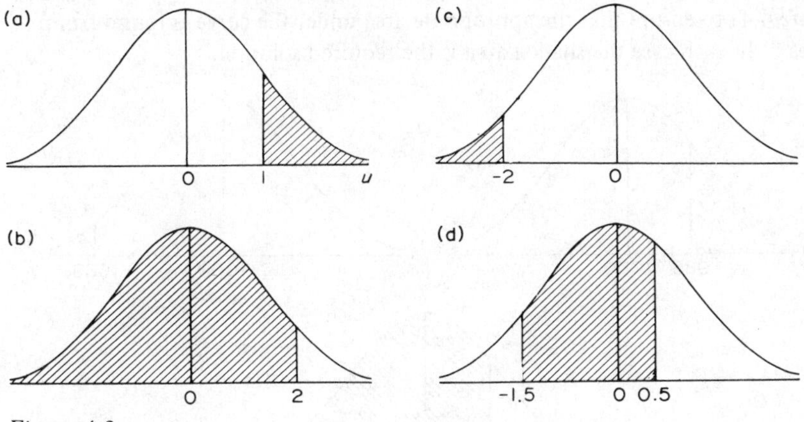

Figure 4.3

(*a*) $u = 1.0$
Area = 0.1587

(*b*) $u = 2.0$
Area in right tail = 0.02275
Thus shaded area = 1 − 0.02275 = 0.97725

(*c*) By symmetry area to left of $u = -2$ is the same as the area to the right of $u = +2$.
Thus the shaded area = 0.02275

(*d*) Area above $u = +0.5$ is 0.3085
Area below $u = -1.5$ is 0.0668

Total unshaded area = 0.3753
∴ shaded area = 0.6247

2. Jam is packed in tins of nominal net weight 1 kg. The actual weight of jam delivered to a tin by the filling machine is normally distributed about the set weight with standard deviation of 12 g.

(*a*) If the set, or average, filling of jam is 1 kg what proportion of tins contain

(i) less than 985 g?
(ii) more than 1030 g?
(iii) between 985 and 1030 g?

(*b*) If not more than one tin in 100 is to contain less than the advertised net weight, what must be the minimum setting of the filling machine in order to achieve this requirement?

(*a*) In solving such problems as these, it is always useful to draw a sketch (figure 4.4) to ensure that the appropriate area under the curve is found from tables.* In each case the shaded area is the required solution.

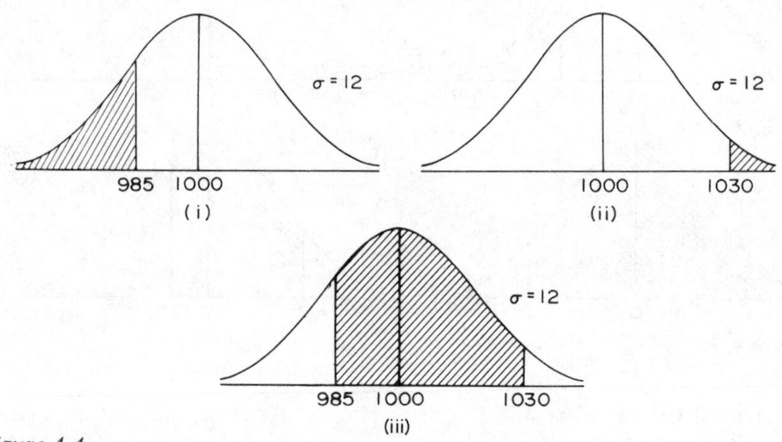

Figure 4.4

(i) $u = \dfrac{985 - 1000}{12} = -1.25$

Using table 3* and the symmetry of the curve, the required proportion is 0.1056

(ii) $u = \dfrac{1030 - 1000}{12} = \dfrac{30}{12} = 2.5$

This corresponds to a right-hand tail area of 0.00621

(iii) To find a shaded area as in this case, the tail areas are found directly from tables* and then subtracted from the total curve area (unity).

The lower and upper tail areas have already been found in (i) and (ii) and thus the solution is

$$1-(0.1056+0.00621)=1-0.1118=0.8882$$

(*b*) In this case, the area in the tail is fixed and in order to find the value of the mean corresponding to this area, the cut-off point (1000 g) must be expressed in terms of the number of standard deviations that it lies from the mean.

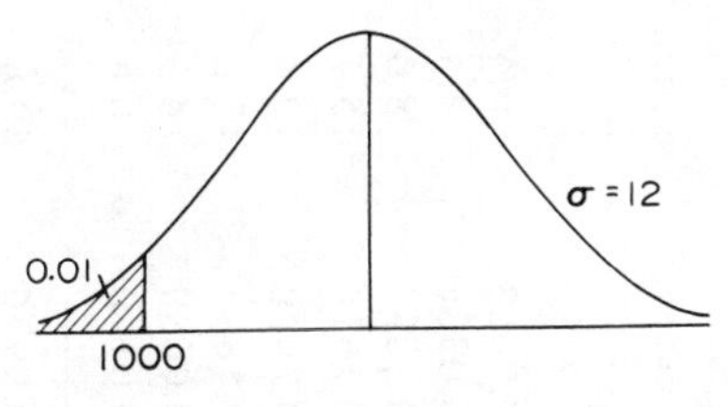

Figure 4.5

From table 4* (or table 3* working from the body of the table outwards), 1% of a normal distribution is cut off beyond 2.33 standard deviations from the mean.

The required minimum value for the mean is thus

$$1000+2.33\times 12=1028\text{ g}=1.028\text{ kg}$$

3. The data from problem 1, chapter 2 (page 46), can be used to show the fitting of a normal distribution. The observed and fitted distributions are also shown plotted on arithmetic probability paper.

The mean of the distribution was 0.087 min and the standard deviation 0.013 min. The method of finding the proportion falling in each class of a normal distribution with these parameters is shown in table 4.1. The expected class frequencies are found by multiplying each class proportion by the total observed frequency. Notice that the total of the expected normal frequencies is not 60. The reason is that about $\frac{1}{4}$% of the fitted distribution lies outside the range (0.045 to 0.125) that has been considered.

Table 4.2 shows the observed and expected normal class frequencies in cumulative form as a percentage of the total frequency. Figure 4.6 shows these two sets of data superimposed on the same piece of normal (or arithmetic) probability paper.

The dots in figure 4.6 represent the observed points and the crosses represent the fitted normal frequencies. Note that the plot of the cumulative normal percentage frequencies does not quite give a straight line. The reason for this is that the $\frac{1}{16}$% of the normal distribution having values less than 0.045 has not been included. If this $\frac{1}{16}$% were added to each of the cumulative percentages in the right-hand column of table 4.2 then a straight-line plot would be obtained.

Class	Class boundary	Standardised upper boundary 'u'	Upper tail area of N.D. above 'u'	Area in each class	Expected normal frequency	Observed class frequency
0.035–0.045	0.045	−3.23	1−0.0006 = 0.9994			
0.045–0.055	0.055	−2.46	1−0.0069 = 0.9931	0.0063	0.4	0
0.055–0.065	0.065	−1.69	1−0.0455 = 0.9545	0.0386	2.3	5
0.065–0.075	0.075	−0.92	1−0.1788 = 0.8212	0.1333	8.0	4
0.075–0.085	0.085	−0.15	1−0.4404 = 0.5596	0.2616	15.7	14
0.085–0.095	0.095	0.62	0.2676	0.2920	17.5	23
0.095–0.105	0.105	1.38	0.0838	0.1838	11.0	9
0.105–0.115	0.115	2.15	0.0158	0.0680	4.1	5
0.115–0.125	0.125	2.92	0.0018	0.0140	0.8	0
				0.9976	59.8	60

Table 4.1

Class	Observed			Fitted normal		
	Frequency	Cumulative frequency	% Cumulative frequency	Frequency	Cumulative frequency	% Cumulative frequency
0.045–0.055				0.4	0.4	0.7
0.055–0.065	5	5	8.3	2.3	2.7	4.5
0.065–0.075	4	9	15.0	8.0	10.7	17.8
0.075–0.085	14	23	38.3	15.7	26.4	44.0
0.085–0.095	23	46	76.7	17.5	43.9	73.2
0.095–0.105	9	55	91.7	11.0	54.9	91.5
0.105–0.115	5	60	100.0	4.1	59.0	98.3
0.115–0.125				0.8	59.8	99.7

Table 4.2

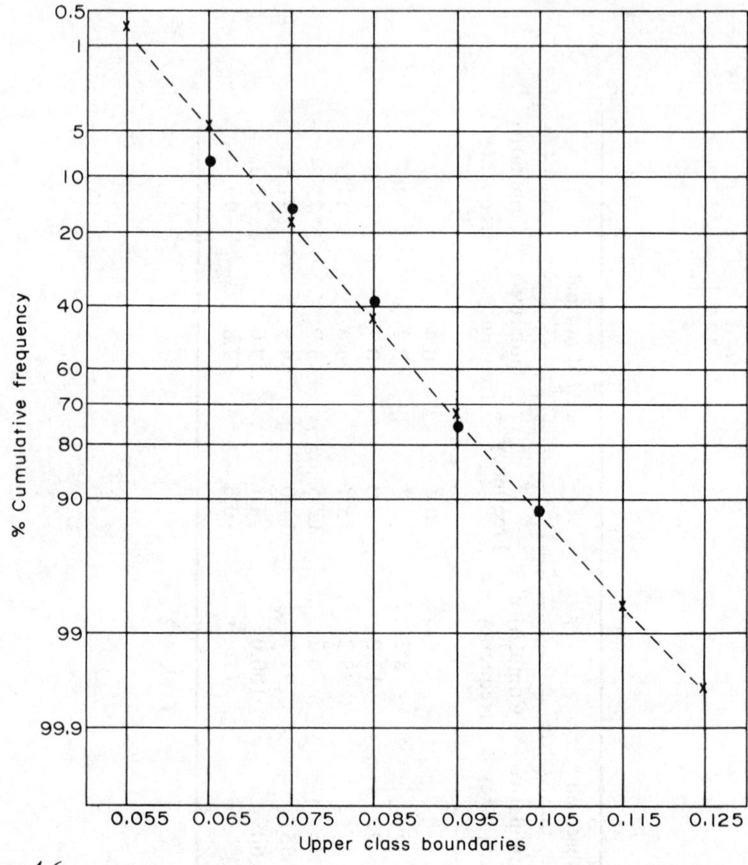

Figure 4.6

A further point to note is that the cumulative frequences are plotted against the upper class boundaries (not the mid point of the class) since those are the values below which lie the appropriate cumulative frequencies.

In addition, if the plotted points fall near enough on a straight line, which implies approximate normality of the distribution, the mean and standard deviation can be estimated graphically from the plot. To do this the best straight line is drawn through the points (by eye is good enough). This straight line will intersect the 16%, 50% and 84% lines on the frequency scale at three points on the scale of the variable.

The value of the variable corresponding to the 50% point gives an estimation of the median, which is the same as the mean if the distribution being plotted is approximately symmetrical.

The horizontal separation between the 84% and 16% intercepts is equal to

2σ for a straight line (normal) plot and so half of this distance gives an estimate of the standard deviation.

Applying this to the fitted normal points, the mean is estimated as 0.087 and the standard deviation comes out as 0.5 (0.100–0.074) = 0.013, the figures used to derive the fitted frequencies in the first place. The small bias referred to earlier caused by omitting the bottom $\frac{1}{16}$% of the distribution in the plot has had very little influence on the estimate in this case.

4.3 Problems for Solution

(** denotes more difficult problems)

1. For any normal distribution, what proportion of it is

(*a*) more than twice the standard deviation above the mean?
(*b*) further than half the standard deviation below the mean?
(*c*) within one and a half standard deviations of the mean?

2. A normal distribution has a mean of 56 and a standard deviation of 10. What proportion of it

(*a*) exceeds 68?
(*b*) is less than 40?
(*c*) is contained between 56 and 65?
(*d*) is contained between 60 and 65?
(*e*) is contained between 52 and 65?

3. Problem 8 of chapter 2 (page 48) gives the intelligence quotients of a sample of 100 children. The mean and standard deviation of these numbers are 99.3 and 13.4, respectively, and the histogram indicates that normality is a good assumption for the distribution of intelligence quotient (I.Q.).

(*a*) What proportion of all children can be expected to have I.Q's

(i) greater than 120?
(ii) less than 90?
(iii) between 70 and 130?

(*b*) What I.Q. will be exceeded by

(i) 1% of children?
(ii) 0.1% of children?
(iii) 90% of children?

(*c*) Between what limits will 95% of children's I.Q. values lie?

What assumptions have been made in obtaining these answers?

4. A process of knitting stockings should give a mean part-finished stocking length of 1.45 m with a standard deviation of 0.013 m. Assuming that the distribution of length is normal,

(*a*) if a tolerance of 1.45 m ± 0.020 m is fixed, what total percentage of oversize and undersize stockings can be expected?

(*b*) What tolerance can be worked to if not more than a total of 5% of stockings undersized or oversized can be accepted?

(*c*) if the mean part-finished length is actually 1.46 m, what proportion of the output are undersized or oversized stockings, allowing a tolerance of 1.45 m ± 0.025 m.

5. The door frames used in an industrialised building system are of one standard size. If the heights of adults are normally distributed, men with a mean of 1.73 m and standard deviation of 0.064 m and women with a mean of 1.67 m and standard deviation of 0.050 m,

(*a*) what proportion of men will be taller than the door frames if the standard frame height is 1.83 m?

(*b*) what proportion of women will be taller than the standard frame height of 1.83 m?

(*c*) what proportion of men will have a clearance of at least 13 cm on a frame height of 1.83 m?

(*d*) what should the minimum frame height be such that at most one man in a thousand will be taller than the frame height?

(*e*) if women outnumber men (e.g. in a large department store) in the ratio 19 : 1, for what proportion of people would a frame height of 1.83 m be too low?

6. The data summarised in table 4.3 come from the analysis of 53 samples of rock taken every few feet during a tin-mining operation. The original data for each sample were obtained in terms of pounds of tin per ton of host rock but since the distribution of such a measurement from point to point is quite skew, the data were transformed by taking the ordinary logarithms of each sample value and summarising the 53 numbers so obtained into the given frequency distribution.
Fit a normal distribution to the data.

**7. The individual links used in making chains have a normal distribution of strength with mean of 1000 kg and standard deviation of 50 kg.
If chains are made up of 20 randomly chosen links

(*a*) what is the probability that such a chain will fail to support a load of 900 kg?

Logarithm of ore grade	Frequency of given ore grade
0.6–0.799	1
0.8–0.999	3
1.0–1.199	6
1.2–1.399	8
1.4–1.599	12
1.6–1.799	11
1.8–1.999	6
2.0–2.199	4
2.2–2.399	2
	53

Table 4.3

(*b*) what should the minimum mean link strength be for 99.9% of all chains to support a load of 900 kg?

(*c*) what is the median strength of a chain?

**8. The standardised normal variate, u, having mean of 0 and variance of 1, has probability density function

$$\phi(u) = \frac{1}{\sqrt{(2\pi)}}\, e^{-\frac{1}{2}u^2}, \qquad -\infty < u < \infty$$

If this distribution is truncated at the point u_α (i.e. the shaded portion, α, of the distribution above u_α is removed—see figure 4.7), obtain an expression in terms of α and u_α showing the amount by which the mean of the truncated distribution is displaced from $u = 0$.

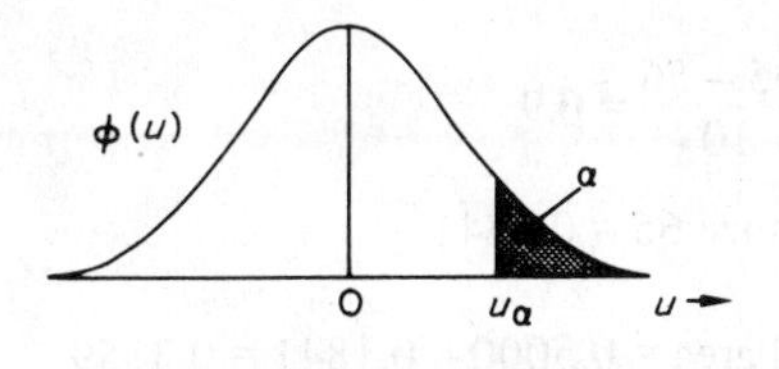

Figure 4.7

9. In a bottle-filling process, the volume of liquid delivered to a bottle is normally distributed with mean and standard deviation of 1 litre and 5 ml respectively. If all bottles containing less than 991 ml are removed and emptied, and the contents used again in the filling process, what will be the average volume of liquid in bottles offered for sale?

4.4 Solutions to Problems

1. Use table 3* of statistical tables.

(*a*) The proportion two standard deviations is 0.02275 (from the table).

(*b*) From the symmetry of the normal distribution, 0.3085 of the area is further than 0.5 standard deviations below the mean.

(*c*) 0.0668 of the distribution is beyond one and a half standard deviations from the mean in each tail. Thus the proportion within 1.5 standard deviations is

$$1-(0.0668+0.0668) = 0.8664$$

2. (*a*) $u = \dfrac{68-56}{10} = \dfrac{12}{10} = 1.2$

Thus, 0.1151 of the area exceeds 68

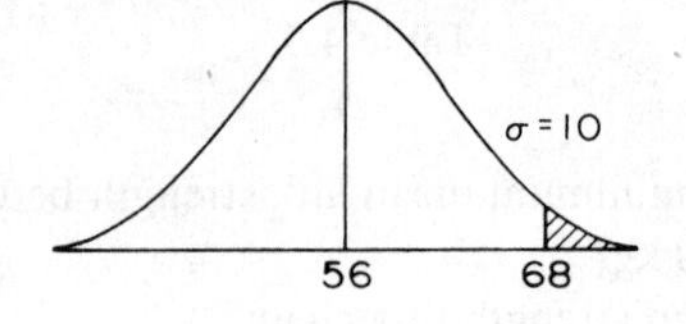

Figure 4.8

(*b*) $u = \dfrac{40-56}{10} = \dfrac{-16}{10} = -1.6$

Thus, 0.0548 of the distribution takes values less than 40.

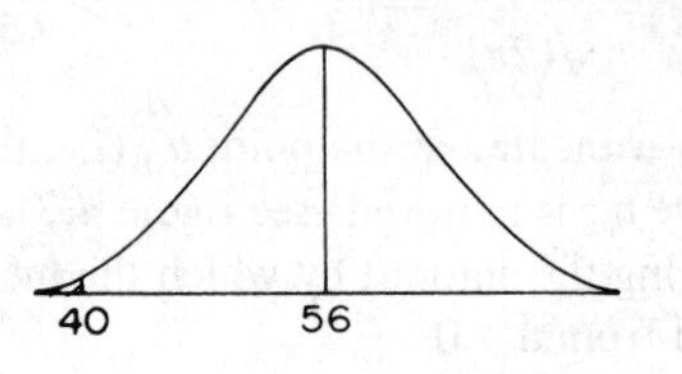

Figure 4.9

(*c*) For 65, $u = \dfrac{65-56}{10} = 0.9$

Area in upper tail above 65 = 0.1841
For 56, $u = 0$
∴ Required shaded area = 0.5000 – 0.1841 = 0.3159

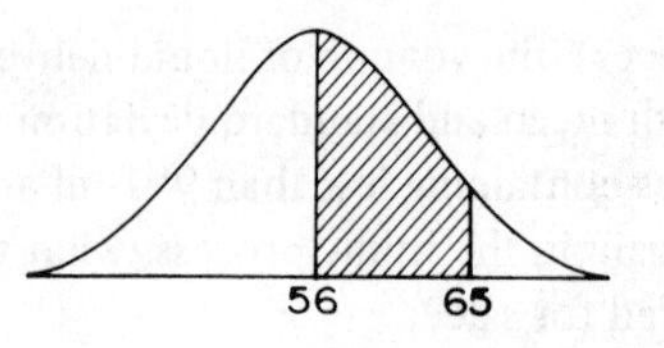

Figure 4.10

(*d*) For 60, $u = \dfrac{60 - 56}{10} = 0.4$

Thus, area above 60 is 0.3446. Area above 65 is found in (*c*) to be 0.1841. Thus, proportion between 60 and 65 is 0.3446 − 0.1841 = 0.1605.

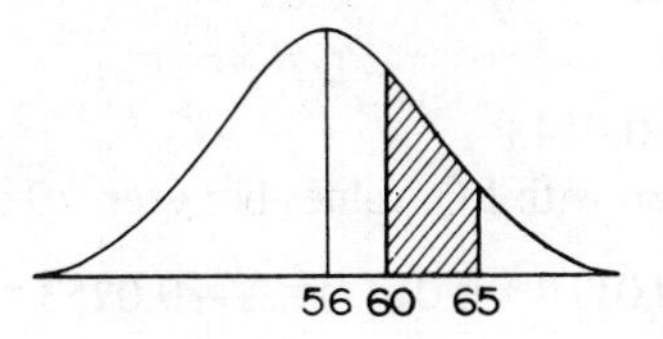

Figure 4.11

(*e*) For 52, $u = \dfrac{52 - 56}{10} = -0.4$

From symmetry, area below 52 = 0.3446.

From (*c*) area above 65 = 0.1841. Thus, proportion between 52 and 65 = 1 − (0.3446 + 0.1841) = 1 − 0.5287 = 0.4713

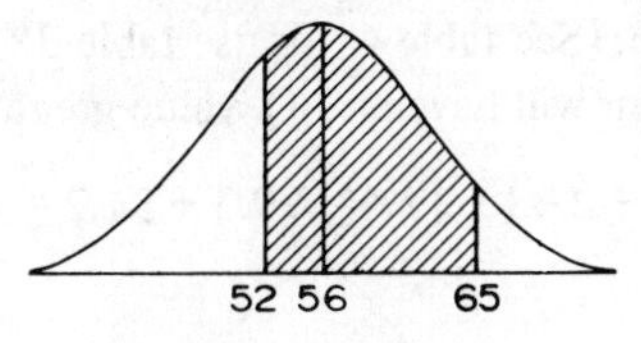

Figure 4.12

3. (*a*) (i) For I.Q. = 120, $u = \dfrac{120 - 99.3}{13.4} = 1.54$

Proportion greater than 120 is 0.0618, say, 0.06

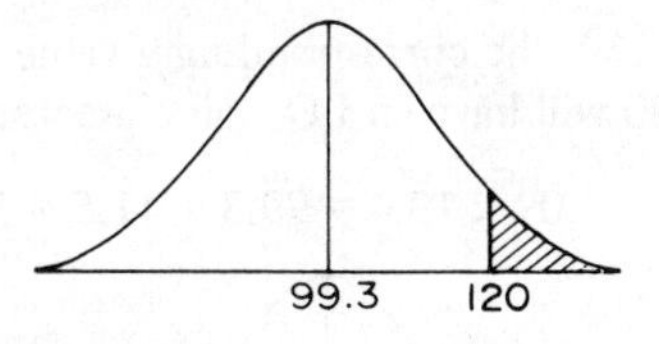

Figure 4.13

(ii) I.Q. = 90, $u = \dfrac{90 - 99.3}{13.4} = -0.69$

By symmetry, proportion less than 90 = 0.2451, say 0.24, since u is nearer to −0.694

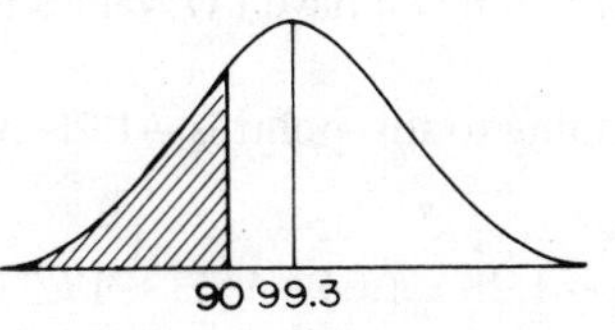

Figure 4.14

(iii) I.Q. = 130, $u = \dfrac{130 - 99.3}{13.4} = 2.29$

Area above $u = 2.29$ is 0.0110.

I.Q. = 70, $u = \dfrac{70 - 99.3}{13.4} = -2.19$

Area below $u = -2.19$ is 0.0143

$\therefore$ Proportion of children with I.Q. values between 70 and 130 is

$$1 - (0.0110 + 0.0143) = 1 - 0.0253 = 0.975$$

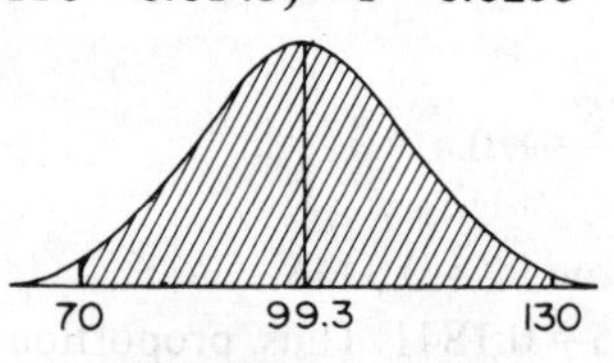

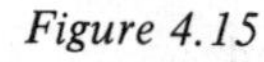
Figure 4.15

(*b*) (i) For all normal distributions, 1% in the tail occurs at a point 2.33 standard deviations from the mean. (See table 4* or use table 3* in reverse.)

Thus, 1% of all children will have an I.Q. value greater than

$$99.3 + 2.33 \times 13.4 = 99.3 + 31.2 = 130.5$$

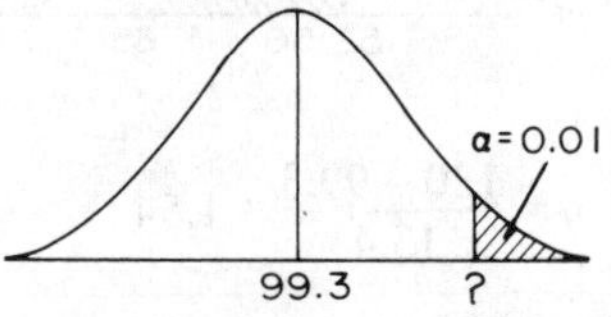

Figure 4.16

(ii) For $\alpha = 0.001$ (0.1%), the corresponding u-value is 3.09.

Thus one child in 1000 will have an I.Q. value greater than

$$99.3 + 3.09 \times 13.4 = 99.3 + 41.5 = 140.8$$

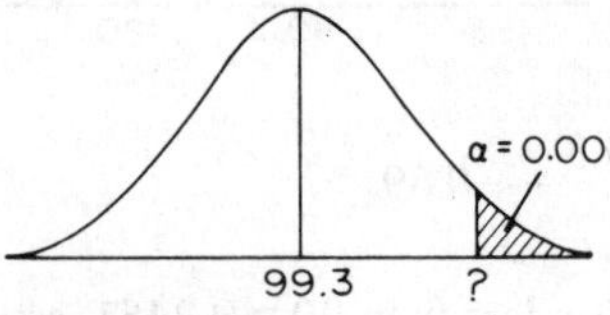

Figure 4.17

(iii) Ten per cent of children will have I.Q. values less than the value which 90% exceed.

The u-value corresponding to this point is -1.28 and converting this into the scale of I.Q. gives

$$99.3 - 1.28 \times 13.4 = 99.3 - 17.2 = 82.1$$

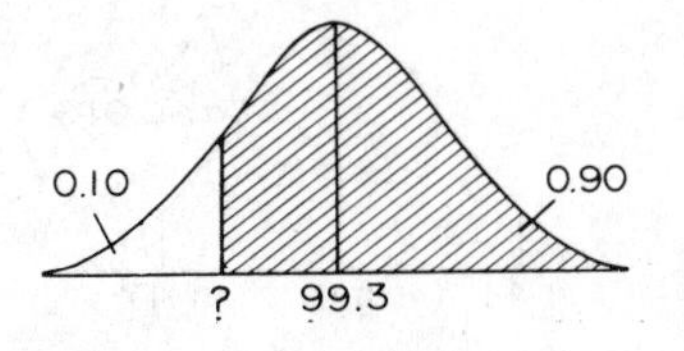

Figure 4.18

(c) We need to find the lower and upper limits such that the shaded area is 95% of the total. There are a number of ways of doing this, depending on how the remaining 5% is split between the two tails of the distribution. It is usual to divide them equally. On this basis, each tail will contain 0.025 of the total area and here the required limits will be 1.96 standard deviations below and above the mean respectively.

Thus, 95% of children will have I.Q. values between

$$99.3 - 1.96 \times 13.4 \quad \text{and} \quad 99.3 + 1.96 \times 13.4 \quad \text{i.e.}$$

$$99.3 - 26.2 \quad \text{and} \quad 99.3 + 26.2$$

$$73.1 \quad \text{and} \quad 125.5$$

Figure 4.19

We have assumed that the original sample of 100 children was taken randomly and representatively from the whole population of children about whom the above probability statements have been made. This kind of assumption should always be carefully checked for validity in practice.

In addition, the mean and standard deviation of the sample were used as though they were the corresponding values for the *population.* In general, they will not be numerically equal, even for samples as large as 100, and this will introduce errors into the statements made. However, the answers will be of the right order of magnitude which is mostly all that is required in practice.

The assumption of normality of the population has already been mentioned.

4. (a) If the mean length is 1.45 m then the maximum deviation allowed for a stocking to be acceptable is

$$\pm \frac{0.020}{0.013}$$

standard deviations, i.e. $u = \pm 1.54$.

The percentage of unacceptable output is represented by the two shaded areas in figure 4.20 and is $2 \times 0.0618 \times 100 = 12.36\%$.

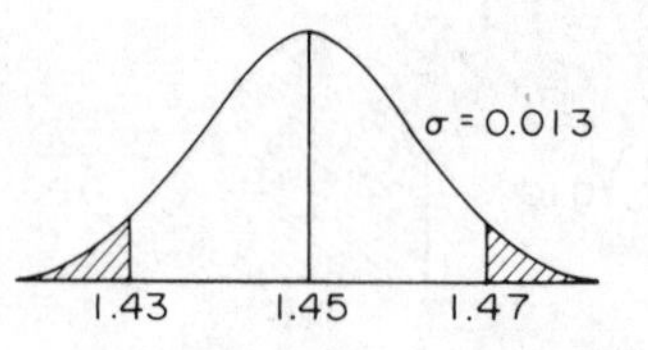

Figure 4.20

(*b*) This time the two shaded areas are each specified to be 0.025 ($2\frac{1}{2}\%$).
Therefore the tolerance that can be worked to corresponds to $u = \pm 1.96$, i.e. to $\pm 1.96 \times 0.013 = \pm 0.025$ m, or ± 25 mm.

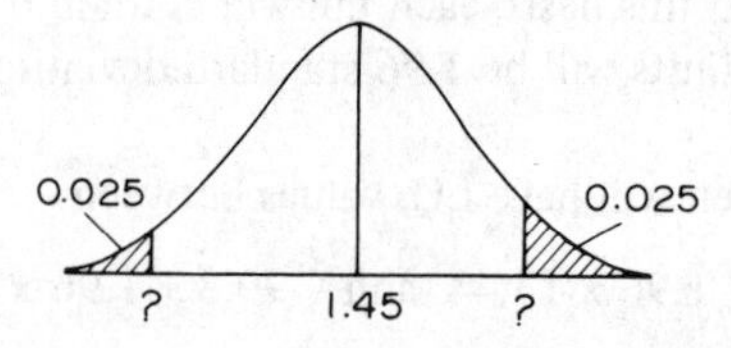

Figure 4.21

(*c*) The lower and upper lengths allowed are 1.425 m and 1.475 m respectively. The shaded area gives the proportion of stockings that do not meet the standard when the process mean length is 1.46 m.

$$1.475 \text{ m}, \quad u = \frac{1.475 - 1.460}{0.013} = 1.15; \qquad \text{area} = 0.1251$$

$$1.425 \text{ m}, \quad u = \frac{1.425 - 1.460}{0.013} = -2.69; \qquad \text{area} = 0.0036$$

Total shaded area = 0.1287

Thus nearly 13% of output will not meet the standard.

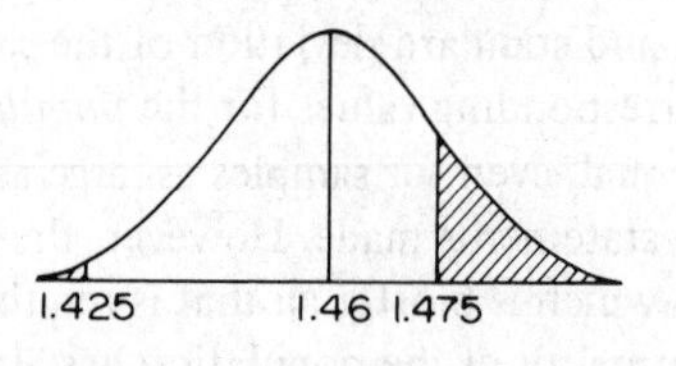

Figure 4.22

5. (*a*) For 1.83 m, $u = \dfrac{1.83 - 1.73}{0.064} = 1.56$

Required proportion = 0.0594

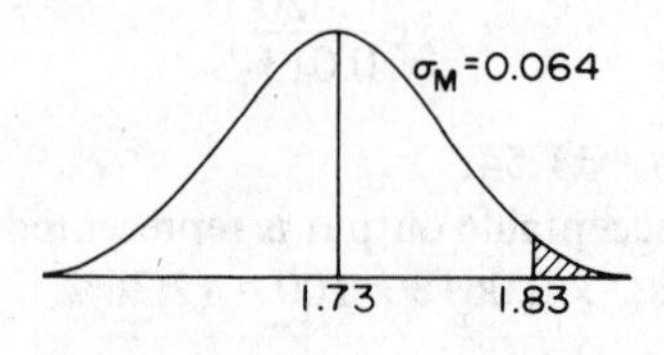

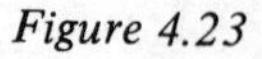

Figure 4.23

(*b*) For 1.83 m, $u = \dfrac{1.83 - 1.67}{0.050} = 3.2.$

Required proportion = 0.00069

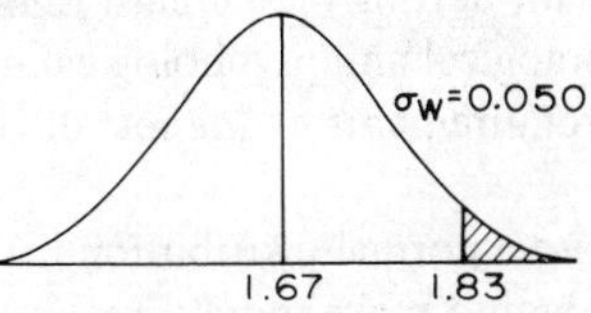

Figure 4.24

(*c*) Men shorter than 1.83 – 0.13 = 1.70 will have a clearance of at least 0.13 m.

Corresponding $u = \dfrac{1.70 - 1.73}{0.064} = -0.47$

From symmetry, proportion of men with at least 13 cm to spare is 0.3192.

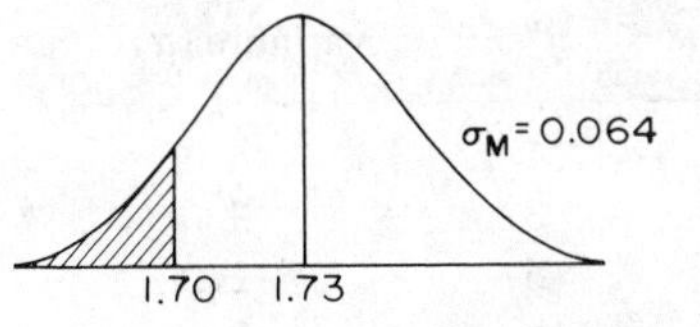

Figure 4.25

(*d*) The frame height which is exceeded by one man in a thousand will be 3.09 standard deviations above the mean height of men, i.e. at

$$1.73 + 3.09 \times 0.064 = 1.93 \text{ m}$$

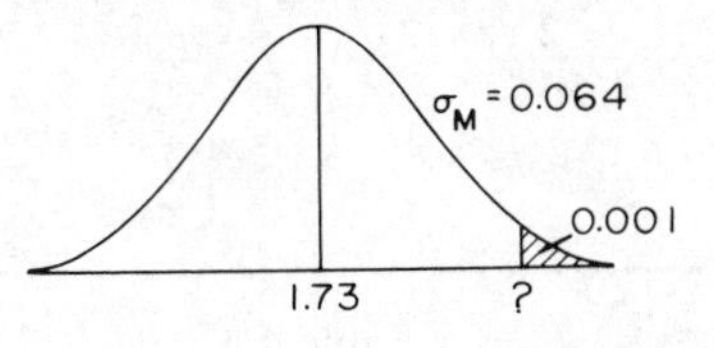

Figure 4.26

(*e*) For women, 1.83 m corresponds to $u = \dfrac{1.83 - 1.67}{0.050} = 3.2$

Proportion of women taller than 1.83 m = 0.00069

For men, 1.83 m corresponds to $u = \dfrac{1.83 - 1.73}{0.064} = 1.56$

Proportion of men taller than 1.83 m = 0.0594

∴ Expected proportion of *people* for whom 1.83 m is too low is

$$0.00069 \times 0.95 + 0.0594 \times 0.05 = 0.004, \text{ i.e}$$

4 people in a 1000.

The problem can be extended by allowing some people to wear hats as well as shoes with different heights of heel.

This problem was intended to give practice in using normal tables of area. Any practical consideration of the setting of standard frame heights would need to take account of the physiological and psychological needs of human door users, of economics and of the requirements of the rest of the building system.

6. It is quite possible to use a normal distribution having an arbitrary mean and standard deviation, but it would make more sense in this case to use the mean and standard deviation of the observed data. The reason for this is that we are mainly concerned with testing the assumption of normality without wishing to specify the parameters.

First the mean and standard deviation are found.

x	f	Coded variable (u)	fu	fu^2
0.6–0.799	1	−4	−4	16
0.8–0.999	3	−3	−9	27
1.0–1.199	6	−2	−12	24
1.2–1.399	8	−1	−8	8
1.4–1.599	12	0	−33	0
1.6–1.799	11	1	11	11
1.8–1.999	6	2	12	24
2.0–2.199	4	3	12	36
2.2–2.399	2	4	8	32
	53		43	178
			−33	
			10	

Table 4.4

$$\text{Mean} = 1.5 + 0.2 \times \tfrac{10}{53} = 1.538$$

$$\text{Standard deviation} = 0.2\sqrt{\left(\frac{178 - \frac{10^2}{53}}{53}\right)} = 0.2\sqrt{\left(\frac{176.1}{53}\right)} = 0.364$$

Using these two values, the areas under the fitted normal curve falling in each class are found using table 3* of the statistical tables. This operation is carried out in table 4.5. Note that the symbol u in the table refers to the standardised normal variate corresponding to the class boundary, whereas in table 4.4 it represents the coded variable (formed for ease of computation) obtained by subtracting 1.5 from each class midpoint and dividing the result by 0.2, the class width.

Class	Class boundaries	u	Area above u	Area in each class	Expected normal frequency
0.4–0.6					
	0.6	−2.58	1 − 0.0049 = 0.9951		
0.6–0.8				0.0163	0.86
	0.8	−2.03	1 − 0.0212 = 0.9788		
0.8–1.0				0.0482	2.55
	1.0	−1.48	1 − 0.0694 = 0.9306		
1.0–1.2				0.1068	5.66
	1.2	−0.93	1 − 0.1762 = 0.8238		
1.2–1.4				0.1758	9.32
	1.4	−0.38	1 − 0.3520 = 0.6480		
1.4–1.6				0.2155	11.42
	1.6	0.17	0.4325		
1.6–1.8				0.1967	10.43
	1.8	0.72	0.2358		
1.8–2.0				0.1338	7.09
	2.0	1.27	0.1020		
2.0–2.2				0.0676	3.58
	2.2	1.82	0.0344		
2.2–2.4				0.0255	1.35
	2.4	2.37	0.0089		
2.4–2.6					

Table 4.5

7. (*a*) Since a chain is as strong as its weakest link, the chain will fail to support a load of 900 kg if *one or more* of its links is weaker than 900 kg.

The probability that a single link is weaker than 900 kg is given by the area in the tail of the normal curve below

$$u = \frac{900 - 1000}{50} = -2, \quad \text{i.e.} \quad 0.02275$$

∴ The probability that a single link does *not* fail at 900 kg = 0.97725 and the probability that none of the links fails = 0.97725^{20}. Thus the probability that a chain of 20 links will not support a load of 900 kg is

$$1 - (0.97725)^{20} = 1 - 0.631 = 0.37$$

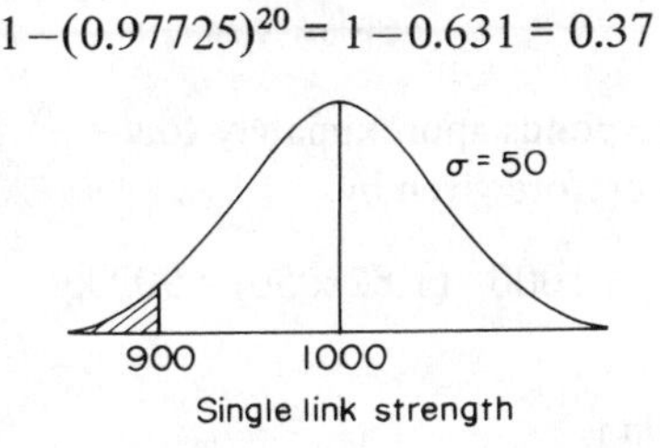

Figure 4.27

(*b*) In this case, the probability of a chain supporting a load of 900 kg is required to be 0.999.

Let p be the probability that an individual link is stronger than 900 kg. Then we have that

$$p^{20} = 0.999$$

$$p = 0.99998 \text{ (using 5 figure logarithms)}$$

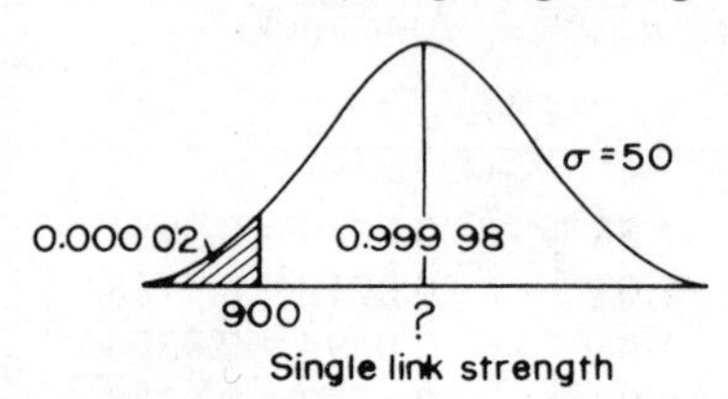

Figure 4.28

It follows that the probability of an individual link's being weaker than 900 kg must be at most 0.00002.

Thus 900 kg corresponds to $u = -4.0$ approximately and the *mean* link strength must be at least

$$900 + 4.0 \times 50 = 1100 \text{ kg}$$

(*c*) In the long run, one chain out of every two will be stronger than the median chain strength.

Let p be the probability that an individual link exceeds the median *chain* strength.

Then from $p^{20} = 0.5$

$$p = 0.96594 \text{ (using 5 figure logarithms)}$$

and the probability that an individual link is *less* than the median chain strength is $(1-p) = 0.0341$.

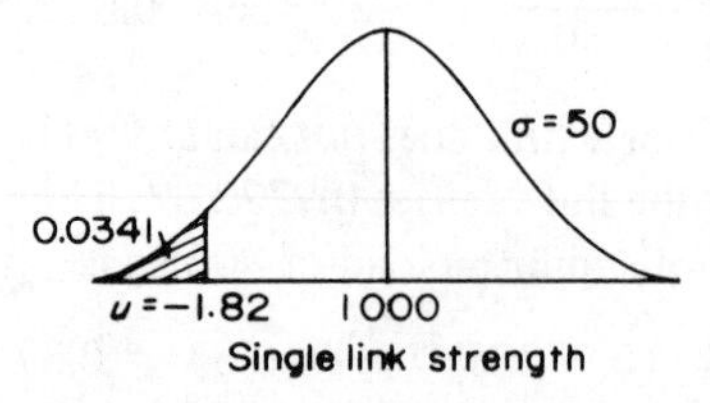

Figure 4.29

Such a tail area corresponds approximately to $u = -1.82$ and the median strength of a chain is therefore given by

$$1000 - (1.82 \times 50) = 909 \text{ kg}$$

8. The density function is

$$\phi(u) = \frac{1}{\sqrt{(2\pi)}} e^{-\frac{1}{2}u^2}$$

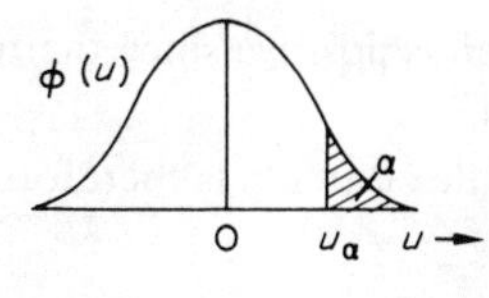

Figure 4.30

The mean of the truncated distribution is given by

$$\frac{\int_{-\infty}^{u_\alpha} u\,\phi(u)\,\mathrm{d}u}{\int_{-\infty}^{u_\alpha} \phi(u)\,\mathrm{d}u} = \frac{1}{\sqrt{(2\pi)}}\,\frac{\int_{-\infty}^{u_\alpha} u\,\mathrm{e}^{-\frac{1}{2}u^2}\,\mathrm{d}u}{(1-\alpha)} = \frac{1}{(1-\alpha)\sqrt{(2\pi)}}\left[-\mathrm{e}^{-\frac{1}{2}u^2}\right]_{-\infty}^{u_\infty}$$

$$= \frac{-1}{(1-\alpha)\sqrt{(2\pi)}}\,\mathrm{e}^{-\frac{1}{2}u_\alpha^2} = -\frac{1}{(1-\alpha)}\,\phi(u_\alpha)$$

Since the mean was previously at $u = 0$ (i.e. when $\alpha = 0$), the above expression also represents the *shift* in mean.

$\phi(u_\alpha)$ is the ordinate (from table 5* of statistical tables) of the normal distribution corresponding to $u = u_\alpha$.

The result just obtained can be used to solve the numerical part of the problem.

The bottle contents are distributed normally but if the segregation process operates perfectly (which it will not do in practice), the distribution of bottle contents offered for sale will correspond to the unshaded part of figure 4.31.

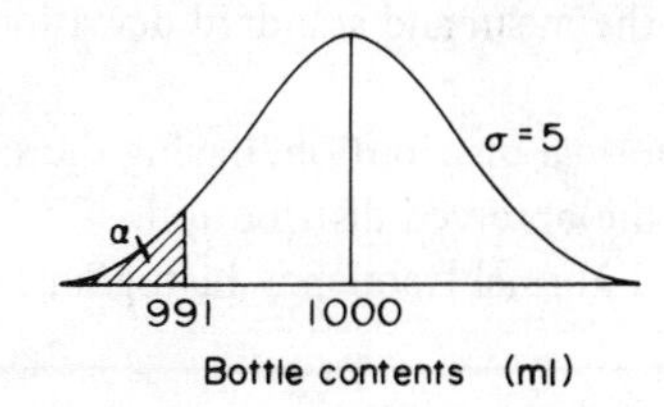

Figure 4.31

The cut-off volume of 991 ml corresponds to

$$u = \frac{991-1000}{5} = 1.8$$

The amount of truncation is therefore $\alpha = 0.0359$. The *increase* in mean volume of despatched bottles is therefore

$$\frac{1}{(1-0.0359)} \times \phi(-1.8) \times \sigma = \frac{0.0790}{0.9641} \times 5 = 0.41 \text{ ml}$$

Note: The change in mean is positive since the truncation occurs in the lower tail instead of the upper tail.

The mean volume of bottle contents is therefore 1000 + 0.41 = 1000.4 ml.

4.5 Practical Laboratory Experiments and Demonstrations

The following experiments are reproduced from *Basic Statistics, Laboratory Instruction Manual*

Appendix 1–Experiment 10

Normal Distribution

Number of persons: 2 or 3.

Object
To give practice in fitting a normal distribution to an observed frequency distribution.

Method
The frequency distribution of total score of three dice obtained by combining all groups' results in table 2, experiment 1, should be re-listed in table 26 (Table 4.6).

Analysis
1. In table 26, calculate the mean and standard deviation of the observed frequency distribution.
2. Using table 27, fit a normal distribution, having the same mean and standard deviation as the data, to the observed distribution.
3. Draw the observed and normal frequency histograms on page 46 and comment on the agreement.

Notes
1. It is not implied in this experiment, that the distribution of the total score of three dice should be normal in form.
2. The total score of three dice is a discrete variable, but the method of fitting a normal distribution is exactly the same for this case as for a frequency distribution of grouped values of a continuous variable.

Class width = unity.

If c = width of class interval, choose x_0 to be the midpoint of a class which, by inspection, is somewhere near the mean of the distribution.

Obtain the class values u from the relation

$$u = \frac{x - x_0}{c}$$

Class Interval	Mid point x	Frequency f	Class u	fu	fu^2
2.5 – 3.5	3				
3.5 – 4.5	4				
4.5 – 5.5	5				
5.5 – 6.5	6				
6.5 – 7.5	7				
7.5 – 8.5	8				
8.5 – 9.5	9				
9.5 – 10.5	10				
10.5 – 11.5	11				
11.5 – 12.5	12				
12.5 – 13.5	13				
13.5 – 14.5	14				
14.5 – 15.5	15				
15.5 – 16.5	16				
16.5 – 17.5	17				
17.5 – 18.5	18				
Totals of +ve terms					
Total of −ve terms					
Net Totals					

The values of u will be positive or negative integers.

The mean, $\bar{x}$, of the sample is given by

$$\bar{x} = x_0 + \frac{\Sigma fu}{\Sigma f}$$

=

= ____________

The variance, $(s')^2$, of the sample is

$$(s')^2 = \left[\frac{\Sigma fu^2 - \dfrac{(\Sigma fu)^2}{\Sigma f}}{\Sigma f}\right]$$

=

= ____________

The standard deviation, s^1, of the sample is given by

$$s^1 = \sqrt{(\text{variance})}$$

=

= ____________

Table 4.6 (Table 26 of the laboratory manual)

Total score of 3 dice	Class boundaries	u for class boundaries	Area under normal curve from u to ∞	Area for each class	Expected normal frequency	Observed frequency
	2.5					
3						
	3.5					
4						
	4.5					
5						
	5.5					
6						
	6.5					
7						
	7.5					
8						
	8.5					
9						
	9.5					
10						
	10.5					
11						
	11.5					
12						
	12.5					
13						
	13.5					
14						
	14.5					
15						
	15.5					
16						
	16.5					
17						
	17.5					
18						
	18.5					

Table 4.7 (Table 27 of the laboratory manual)

Notes

1. u is the deviation from the mean, of the class boundary expressed as a multiple of the standard deviation (with appropriate sign).

$$\text{i.e. } u = \frac{\text{class boundary} - \bar{x}}{s}$$

2. The area under the normal curve above each class boundary may be found from the table of area under the normal curve at the end of the book.

The normal curve area or probability for each class is obtained by differencing the cumulative probabilities in the previous column.

3. Other tables which cumulate the area under the normal curve in a different way may be used, but some of the column headings will require modification and the probabilities subtracted or summed as appropriate.

4. In order to obtain equality of expected and observed total frequencies, the two extreme classes should be treated as open-ended, i.e. with class boundaries of $-\infty$ and $+\infty$ instead of 2.5 and 18.5 respectively.

Appendix 2—Experiment 11

Normal Distribution

Number of persons: 2 or 3.

Object

To calculate the mean and standard deviation of a sample from a normal population and to demonstrate the effect of random sampling fluctuations.

Method

From the red rod population M6/1 (Normally distributed with a mean of 6.0 and standard deviation of 0.2) take a random sample of 50 rods and measure their lengths to the nearest tenth of a unit using the scale provided. The rods should be selected one at a time and replaced after measurement, before the next one is drawn.

Record the measurements in table 28.

Care should be taken to ensure good mixing in order that the sample is random. The rod population should be placed in a box and stirred-up well during sampling.

Analysis

1. Summarise the observations into a frequency distribution using table 29.
2. Calculate the mean and standard deviation of the sample data using table 30.
3. Compare, in table 31, the sample estimates of mean and standard deviation obtained by each group. Observe how the estimates vary about the actual population parameters.
4. Summarise the observed frequencies of all groups in table 32. On page 51, draw, to the same scale, the probability histograms for your own results and for the combined results of all groups. Observe the shapes of the histograms and comment.

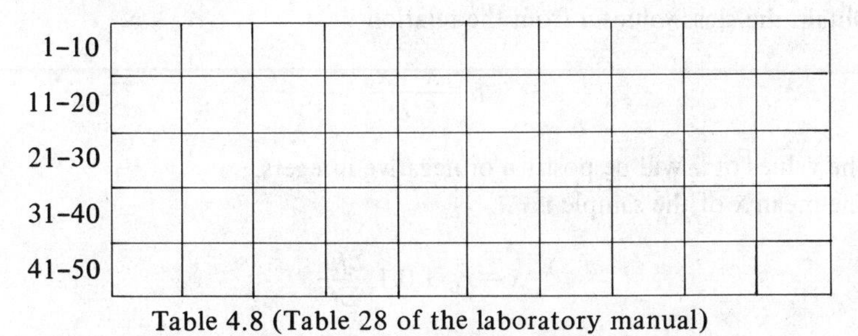

1–10										
11–20										
21–30										
31–40										
41–50										

Table 4.8 (Table 28 of the laboratory manual)

Summarise these observations into class intervals of width 0.1 unit with the measured lengths at the mid points using the 'tally-mark' method and table 29.

Class interval (units)	Class mid point	'Tally-marks'	Frequency
5.35–5.45	5.4		
5.45–5.55	5.5		
5.55–5.65	5.6		
5.65–5.75	5.7		
5.75–5.85	5.8		
5.85–5.95	5.9		
5.95–6.05	6.0		
6.05–6.15	6.1		
6.15–6.25	6.2		
6.25–6.35	6.3		
6.35–6.45	6.4		
6.45–6.55	6.5		
6.55–6.65	6.6		
		Total frequency	

Table 4.9 (Table 29 of the laboratory manual)

Width of class interval 0.1 unit.

If c is width of class interval, choose x_0 to be the mid point of a class which, by inspection, is somewhere near the mean of the distribution.

Obtain the class values u from the relation

$$u = \frac{x - x_0}{c}$$

The values of u will be positive or negative integers.

The mean $\bar{x}$ of the sample is

$$\bar{x} = x_0 + 0.1 \frac{\Sigma fu}{\Sigma f}$$

$$=$$

$$= \underline{\qquad\qquad\qquad}$$

Class Interval, units	Mid point x	Frequency f	Class u	fu	fu^2
5.35–5.45	5.4				
5.45–5.55	5.5				
5.55–5.65	5.6				
5.65–5.75	5.7				
5.75–5.85	5.8				
5.85–5.95	5.9				
5.95–6.05	6.0				
6.05–6.15	6.1				
6.15–6.25	6.2				
6.25–6.35	6.3				
6.35–6.45	6.4				
6.45–6.55	6.5				
6.55–6.65	6.6				
Total of +ve terms					
Total of −ve terms					
Net totals					

Table 4.10 (Table 30 of the laboratory manual)

The variance $(s')^2$ of the sample is

$$(s')^2 = 0.1^2 \left[\frac{\Sigma fu^2 - \frac{(\Sigma fu)^2}{\Sigma f}}{\Sigma f} \right]$$

=

=

The standard deviation s of the sample is given by

$$s' = \sqrt{(\text{variance})}$$

=

=

Group	Sample size	Sample	
		Mean	Standard deviation
1			
2			
3			
4			
5			
6			
7			
8			
Population parameters		6.00	0.2

Table 4.11 (Table 31 of the laboratory manual–summary of data)

	Frequency of rod lengths													
	5.4	5.5	5.6	5.7	5.8	5.9	6.0	6.1	6.2	6.3	6.4	6.5	6.6	
1														
2														
3														
4														
5														
6														
7														
8														
Total frequencies (all groups)														

Table 4.12 (Table 32 of the laboratory manual)

5 Relationship between the basic distributions

5.1 Syllabus Covered

The relationships between hypergeometric, binomial, Poisson and normal distributions; use of binomial as approximation to hypergeometric in sampling; Poisson as approximation to binomial; normal as an approximation to binomial; normal as an approximation to Poisson.

5.2 Résumé of Theory

The following points should be revised and stressed.

(1) The basic laws of the distributions in chapters 3 and 4.

(2) Their interrelationships and conditions for using approximate distributions. Tables 5.1 and 5.2 summarise the interrelationships together with rules for use of the approximations.

Note: In practice use the Poisson and normal distributions as approximations to hypergeometric and binomial whenever possible:

(3) (*a*) Binomial approximation to hypergeometry.
 (*b*) Using Poisson approximation to binomial.
 (*c*) Normal approximation to binomial.
 (*d*) Normal approximation to Poisson.

(4) Whenever the normal distribution is used to approximate either the hypergeometric, binomial or Poisson distributions, care should be taken to remember that a continuous distribution is being approximated to a discrete one, and to include an allowance when calculating probabilities. For example, take the case of using the normal approximation to the following problem.

What is the chance in a group of 100, of more than 20 persons dying before 65 years of age, given that the chance of any one person's dying is 0.20?

Here, since $p > 0.20$ and $np > 5$, normal approximation can be used. However, in calculating the probability figure 5.1 illustrates that the value 20.5 and *not* 20 must be used.

	1 Hypergeometric distribution	2 Binomial distribution	3 Poisson distribution	4 Normal distribution
General term of distribution $P(x)$	$\frac{\binom{M}{x}\binom{N-M}{n-x}}{\binom{N}{n}}$	$\binom{n}{x} p^x(1-p)^{n-x}$	$e^{-m} \frac{m^x}{x!}$	Density function $f(x) = \frac{1}{\sigma\sqrt{2\pi}} e^{\frac{-(x-\mu)^2}{2\sigma^2}}$
Mean	$n \times \frac{M}{N}$	np	m	μ
Variance	$n \times \frac{M}{N}\left(1 - \frac{M}{N}\right)\frac{(N-n)}{(N-1)}$	$np(1-p)$	m	σ^2
Notes:	Computation of probabilities is excessive and the distribution formulae need only be used in practice where the usual approximations (i.e. distribution (2) and its approximations) do not give the required accuracy.	Computation will usually be tedious even if it is practicable. If tables of probabilities are not available, the appropriate one of distributions (3) and (4) can generally be used as a good approximation.	Direct computation is easier than for (1) or (2). Tables of Poisson probabilities are readily available.	Probabilities can easily be obtained from a table of areas under the normal curve. It is necessary to express the variable in standardised form, i.e. in terms of $u = \frac{x-\mu}{\sigma}$

Table 5.1. Relationship between distributions

Use (2) as an approximation for (1)	
putting $p = \dfrac{M}{N}$	if $\dfrac{n}{N} < 0.10$
Use (3) as an approximation for (2)	
putting $m = np$	if $p < 0.10$
Use (4) as an approximation for (2)	if $p \geqslant 0.10$
putting $\mu = np$ and $\sigma^2 = np(1-p)$	$p \leqslant 0.90$ $np > 5$
Use (4) as an approximation for (3)	if $m \geqslant 15$
putting $\mu = m$ and $\sigma^2 = m$	and preferably $m > 30$

Table 5.2 Approximations and a guide to their use in practice

The suggested approximations will usually be satisfactory for practical purposes. However, for values of the parameters near to the limiting conditions give above, care should be taken when determining probabilities in the tails of a distribution, as the errors of approximation may be considerably greater than allowable.

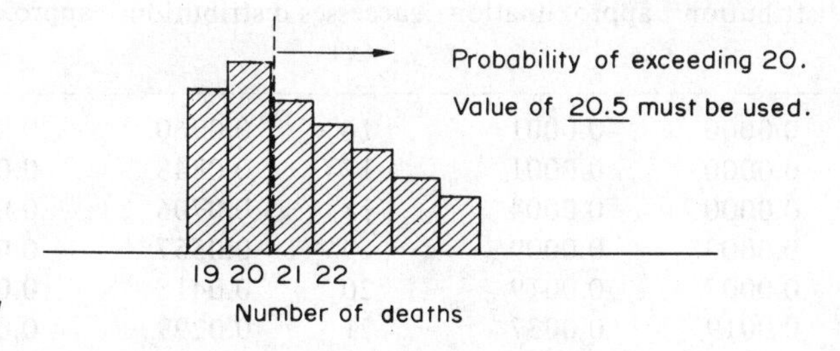

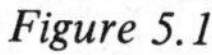
Figure 5.1

5.2.1 *Hypergeometric, Binomial and Poisson Approximations*

Tables 5.3 and 5.4 give details of the accuracy of the approximations at the limiting conditions; obviously the further the parameters are from these conditions the more accurate the approximation.

Batch size $N = 100$ of which 10 are defective.

Sample size $n = 10, \quad p = 0.10$

thus $\dfrac{n}{N} = 0.10, \quad np = 1$

Table 5.3 gives a full comparison of the probabilities of finding x defects in the sample.

No. of defects in sample (x)	Hypergeometric distribution	Binomial distribution	Poisson distribution
0	0.3305	0.3487	0.3679
1	0.4080	0.3874	0.3679
2	0.2015	0.1937	0.1839
3	0.0518	0.0574	0.0613
4	0.0076	0.0112	0.0153
5	0.0006	0.0015	0.0031
6 or over	0.0000	0.0001	0.0006

Table 5.3

5.2.2 *Normal Distribution as an Approximation to Poisson*

Table 5.4 gives a full comparison of the probability of x successes where $m = 15$ using the Poisson and its normal approximation.

No. of successes (x)	Poisson distribution	Normal approximation	No. of successes (x)	Poisson distribution	Normal approximation
0	0.0000	0.0001	16	0.0960	0.0993
1	0.0000	0.0001	17	0.0848	0.0899
2	0.0000	0.0004	18	0.0706	0.0762
3	0.0002	0.0009	19	0.0557	0.0603
4	0.0007	0.0019	20	0.0418	0.0452
5	0.0019	0.0037	21	0.0299	0.0313
6	0.0048	0.0072	22	0.0204	0.0200
7	0.0104	0.0122	23	0.0132	0.0122
8	0.0194	0.0200	24	0.0083	0.0072
9	0.0325	0.0313	25	0.0050	0.0037
10	0.0486	0.0452	26	0.0029	0.0019
11	0.0663	0.0603	27	0.0016	0.0009
12	0.0828	0.0762	28	0.0008	0.0004
13	0.0956	0.0899	29	0.0005	0.0001
14	0.1025	0.0993	30	0.0002	0.0001
15	0.1024	0.1026	31	0.0001	0.0000

Table 5.4

The two distributions converge rapidly as m increases.

While most statisticians accept the use of the normal approximation for $m > 15$, it will be seen that there is quite an appreciable divergence in the tails of the distributions. The authors recommend that whenever possible normal approximation is used when $m > 30$.

The statistical tables* have been amended to give Poisson probabilities up to $m = 40$.

5.2.3 *Examples on the Use of Theory*

1. In sampling from batches of 5000 components, a sample of 50 is taken and if one or more defects is found the batch is rejected. What is the probability of accepting batches containing 2% defects?

The theoretically correct distribution is the hypergeometric, but since

$$\frac{n}{N} \text{ i.e. } \frac{50}{5000}$$

is less than 10%, the binomial can be used.

However, computation is still difficult and since $p < 0.10$ the Poisson distribution can be used.

Solution by binomial approximation from table 1*

Probability of accepting batches with 2% defectives = 0.3642

Solution by Poisson approximation

Expected number of defects in sample = $np = 50 \times 0.02 = 1$

From table 2*

Probability of accepting batches with 2% defectives = 0.3679

2. In 50 tosses of an unbiased coin, what is the probability of more than 30 heads occurring?

This requires the binomial distribution which gives

$$P(>30 \text{ heads}) = \sum_{x=31}^{50} \binom{50}{x} (\tfrac{1}{2})^x (\tfrac{1}{2})^{50-x}$$

from table 1 $= 0.0595$

Using normal approximation since $p > 0.10$ and $np > 5$

mean of distribution = $np = 25$

Variance of distribution = $np(1-p) = 50 \times \frac{1}{2} \times \frac{1}{2} = 12.5$

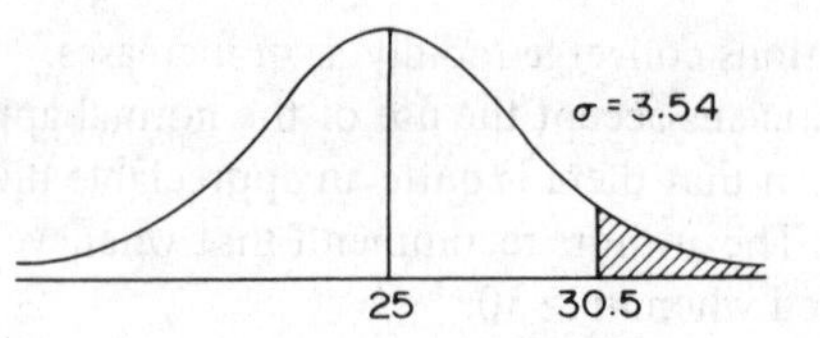

Figure 5.2

Standard deviation = 3.54

$$u = \frac{30.5 - 25}{3.54} = \frac{5.5}{3.54} = 1.55$$

which from table 3* leads to a probability of 0.0606.

Note: Since a continuous distribution is being used to approximate to a discrete distribution, the value 30.5 and *not* 30 must be used in calculating the u value.

3. A machine produces screws 10% of which have defects. What is the probability that, in a sample of 500

(*a*) more than 35 defects are found?
(*b*) between 30 and 35 (inclusive) defects are found?

The binomial law: assuming a sample of 500 from a batch of at least 5000. The normal approximation can be used since $p > 0.10$, $np = 50$.

$$\mu = np = 500 \times 0.10 = 50$$

$$\sigma = \sqrt{(500 \times 0.10 \times 0.90)} = \sqrt{45} = 6.7$$

(*a*) $u = \dfrac{35.5 - 50}{6.7} = -\dfrac{14.5}{6.7} = -2.16$

$\therefore$ Probability of more than 35 defects from tables* = 1 − 0.01539 = 0.9846

(*b*) Probability of between 30 and 35 defects, use limits 29.5 and 35.5.

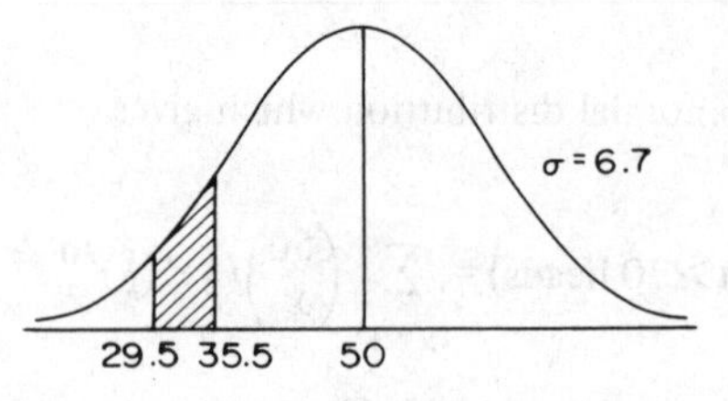

Figure 5.3

$$u = \frac{29.5 - 50}{6.7} = -\frac{20.5}{6.7} = -3.06$$

Probability of more than 29 = 1 − 0.0011 = 0.9989

$\therefore$ Probability of between 30 and 35 inclusive = 0.9989 − 0.9846 = 0.0143

4. The average number of breakdowns per period of an assembly moulding line is 30. If the breakdowns occur at random what is the probability of more than 40 breakdowns occurring per period?

Here the theoretically correct distribution is the Poisson. However, since $m > 15$ use normal approximation.
Solution by Poisson, table 2*

$$P(x) = \sum_{x=41}^{\infty} \frac{30^x e^{-30}}{x!} = 0.0323$$

Using normal approximation

$$\mu = 30 \qquad \sigma^2 = 30$$

$\therefore \quad \sigma = \sqrt{30} = 5.48$

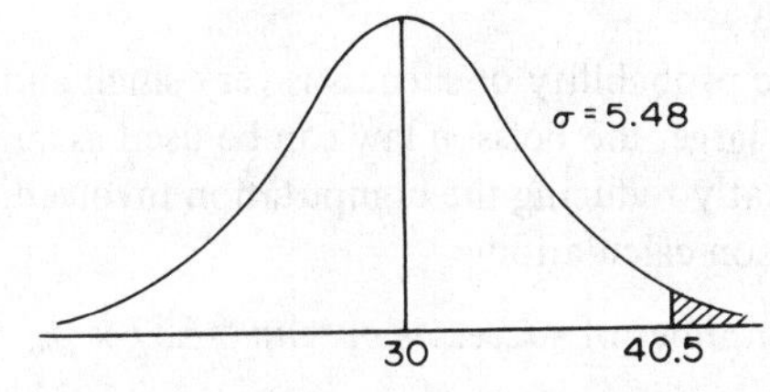

Figure 5.4

Again to include 41 breakdowns but exclude 40

$$u = \frac{40.5 - 30}{5.48} = 1.92$$

Probability of exceeding 40, $P(>40) = 0.0274$ from statistical table 3*.

5.2.4 *Examples of Special Interest*

1. *Bomb Attack on London*

During the last war, it was asked whether the bombs dropped on London were aimed or whether they fell at random. The term 'aimed' is of course very loose, since obviously the Germans could point the bombs towards Britain, but aim in this problem is defined as pinpointing targets inside a given area.

To determine the solution, part of London was divided into 576 equal areas ($\frac{1}{4}$ km^2 each) and the number of areas with 0, 1, 2, . . ., hits was tabulated from the results of 537 bombs which fell on the area. These data in distribution form are shown in table 5.5.

Number of hits j	0	1	2	3	4	5
Number of areas with j hits	229	211	93	35	7	1

Table 5.5

In statistical logic, as will be seen later, an essential step in testing in the logic is the setting up of what is called the *null hypothesis.*

Here the null hypothesis is that the bombs are falling randomly or that there is no ability to aim at targets of the order of $\frac{1}{4}$ km^2 in area.

Then if the hypothesis is true, the probability of any given bomb falling in any one given area = $\frac{1}{576}$.

$\therefore$ Probability of x hits in any area

$$P(x) = \binom{537}{x}\left(\frac{1}{576}\right)^x \left(\frac{575}{576}\right)^{537-x}$$

from the binomial law.

However, since the probability of success is very small and the number of attempts is relatively large, the Poisson law can be used as an approximation to the binomial thus greatly reducing the computation involved.

Thus, for the Poisson calculation

$$\text{average number of successes } m = np = 537 \times \tfrac{1}{576} = 0.93$$

The results obtained by reference to statistical tables by interpolation for the chance of various numbers of hits are given in table 5.6.

Number of hits j	0	1	2	3	4	5
Probability of j hits	0.395	0.367	0.170	0.053	0.012	0.002

Table 5.6

Table 5.7 shows the results obtained by comparing the actual frequency distribution of number of hits per area with the Poisson expected frequencies if the hypothesis is true.

Number of hits j	0	1	2	3	4	5
Actual number of areas with j hits	229	211	93	35	7	1
Expected number of areas with j hits (Poisson)	227	211	98	31	7	1

Table 5.7

The agreement is certainly good enough (without significance testing) to state that the null hypothesis is true; namely, that the bombs fell at random, so that the area into which the bomb could be aimed must have been much larger than the area of London.

2. *Defective Rate of a Production Process*

The number of defects per shift produced by a certain process over the last 52 shifts is given in table 5.8. Is the process in control, i.e. has the defective rate remained constant over the period? The total production per shift is 600 units.

Number of defects/shift	0	1	2	3	4	5	6	7	8	9	10	Total
Frequency	2	6	9	11	8	6	4	3	2	1	0	52

Average defects/shift = 3.6

Table 5.8

This problem gives an excellent introduction to the basic principles of quality control.

The process is assumed to be in control. If this hypothesis is true then

$$\text{the probability of any one component being defective} = \frac{3.6}{600} = 0.006$$

Thus, by the Binomial law

$$\text{probability of } x \text{ defects in a shift } P(x) = \binom{600}{x} 0.006^x\, 0.994^{600-x}$$

Number of defectives(s)	Number of shifts	Poisson $p(s)$	Calculated number of shifts: 52 $P(s)$
0	2	0.0273	1
1	6	0.0984	5
2	9	0.1771	9
3	11	0.2125	11
4	8	0.1912	10
5	6	0.1377	7
6	4	0.0826	4.5
7	3	0.0425	2
8	2	0.0191	1
9	1	0.0076	0.5
10	0	0.0040	0.2
	52	1.0000	51.2

Table 5.9

However, here again the Poisson law gives an excellent approximation to the binomial, reducing the computation considerably.

It should be noted that in most attribute quality control tables this Poisson approximation is used.

Using $m = 3.6$, table 5.9 gives the comparison of the actual pattern of variation with the Poisson.

Reference to the table indicates that the defects in the period of 52 shifts did not show any 'abnormal' deviations from the expected number.

Thus, this comparison gives the basis for determining whether or not a process is in control, the basic first step in any quality control investigation.

5.3 Problems for Solution

1. In a machine shop with 250 machines, the utilisation of each machine is 80%, i.e. 20% of the time the machine is not working. What is the probability of having

   (*a*) more than 60 machines idle at any one time?
   (*b*) between 60 and 65 machines idle?
   (*c*) less than 32 machines idle?

2. In a sampling scheme, a random sample of 500 is taken from each batch of components received. If one or more defects are found the batch is rejected. What is the probability of rejecting batches containing

   (*a*) 1% defectives?
   (*b*) 0.1% defectives?

3. Assuming equal chance of birth of a boy or girl, what is the probability that in a class of 50 students, less than 30% will be boys?

4. The average number of customers entering a supermarket in 1 h is 30. Assuming that all customers arrive independently of each other, what is the probability of more than 40 customers arriving in 1 h?

5. In a hotel, the five public telephones in the lobby are utilised 48% of the time between 6 p.m. and 7 p.m. in the evening. What is the probability of

   (*a*) all telephones being in use?
   (*b*) four telephones being in use?

6. A city corporation has 24 dustcarts for collection of rubbish in the city. Given that the dustcarts are 80% utilised or 20% of time broken down, what proportion of the time will there be more than three dustcarts broken down?

7. A batch of 20 special resistors are delivered to a factory. Four resistors are

defective. Four resistors are selected at random and installed in a control panel. What is the probability that no defective resistor is installed?

5.4 *Worked Solutions to the Problems*

1. This is the binomial distribution.

Since $p = 0.20$, $np = 250 \times 0.20 = 50$, thus normal approximation can be used.

$$\mu = 5.0$$

$$\sigma^2 = 250 \times 0.20 \times 0.80 = 40$$

$\therefore \quad \sigma = 6.3$

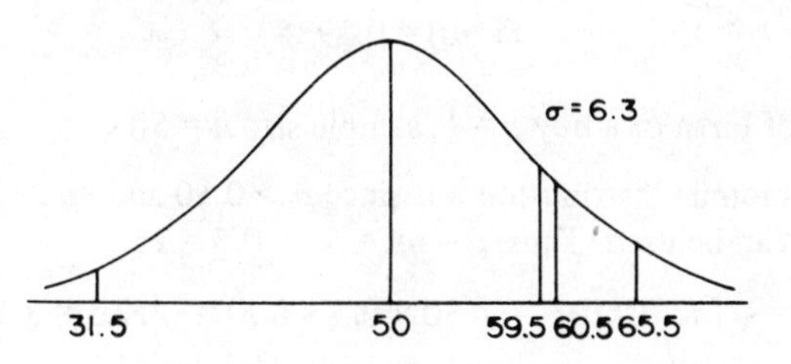

Figure 5.5

(*a*) $u = \dfrac{60.5-50}{6.3} = 1.67$

$\therefore$ Probability of more than 60 machines idle $P(>60) = 0.0475$

(*b*) $u = \dfrac{65.5-50}{6.3} = 2.46$

$\therefore$ Probability of more than 65 machines idle $P(>65) = 0.0069$

Also $u = \dfrac{59.5-50}{6.3} = 1.51$

$\therefore$ Probability of more than 59 machines idle $P(>59) = 0.0655$

$\therefore$ Probability of between 60 and 65 machines idle (inclusive = 0.0655–0.0069

= 0.0586

(*c*) $u = \dfrac{31.5-50}{6.3} = 2.94$

$\therefore$ Probability of less than 32 machines idle $P(<32) = 0.0016$

2. This is the hypergeometric distribution in theory but if it is assumed that n, the sample size, is less than 10% of the batch, the binomial can be used.

For 1% defectives, since $p < 0.10$ and for 0.1% defectives, since $p < 0.10$ } the Poisson approximation can be used.

(*a*) $m = np = 500 \times \frac{1}{100} = 5$

Probability of rejecting batches with 1% defectives

$$P(>0) = 0.9933$$

(*b*) $m = np = 500 \times \frac{0.1}{100} = 0.5$

Probability of rejecting batches with 0.1% defectives

$$P(>0) = 0.3935$$

3. Probability of birth of a boy $p = \frac{1}{2}$, sample size $n = 50$

This is the binomial distribution but since $p > 0.10$ and $np > 5$ the normal approximation can be used. Thus, $\mu = np = 50 \times 0.5 = 25$

$$\sigma = \sqrt{[np(1-p)]} = \sqrt{(50 \times 0.5 \times 0.5)} = \sqrt{12.5} = 3.54$$

To calculate the probability of there being less than 15 boys

$$u = \frac{14.5-25}{3.54} = \frac{-10.5}{3.54} = -2.97$$

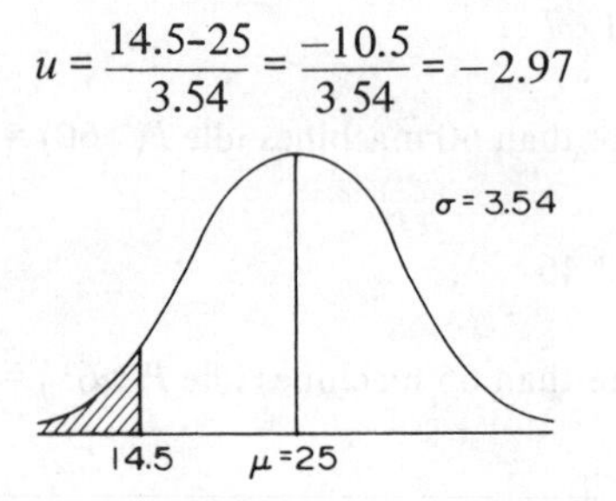

Figure 5.6

∴ From tables*,
Probability of class of 50 having less than 15 boys = 0.0015

Compare this with the correct answer from binomial tables of 0.0013.

4. This by definition is the Poisson law. However, since $m > 15$, the normal approximation can be used. Here $\mu = 30$, $\sigma = \sqrt{30} = 5.48$

$$u = \frac{40.5-30}{5.48} = 1.92$$

∴ Probability of more than 40 customers arriving in 1 h = 0.0274

(Compare this with the theoretically correct result from Poisson of 0.0323.)

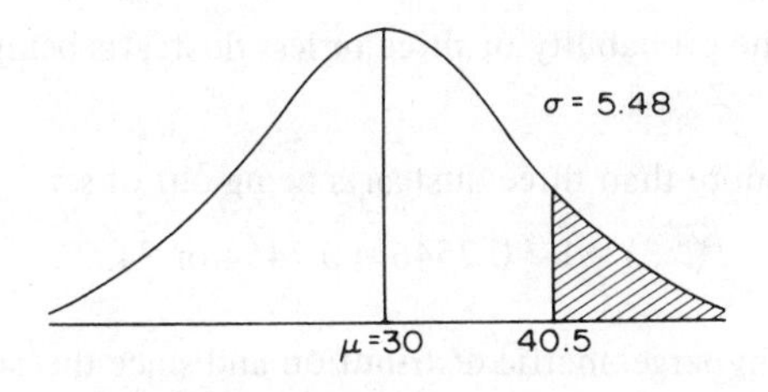

Figure 5.7

5. Probability of a telephone booth being busy $p = 0.48$.

$$\text{number of booths, } n = 5$$

This is the binomial distribution.

This example is included to demonstrate clearly that cases will arise in practice where the approximations given will not apply. Here we cannot use Poisson approximation since $p > 0.10$. Also we cannot use normal approximation since np is not greater than 5. Thus the problem must be solved by computing the binomial distribution or referring to comprehensive binomial tables. Thus

$$\text{probability of all telephones being in use} = 0.48^5 = 0.0255$$

$$\text{probability of four telephones being in use} = \binom{5}{4} 0.48^4 \times 0.52^1 = 0.1380$$

6. Here, $n = 24$

Probability of dustcart's being broken down $(p) = 0.20$. This is the binomial distribution. Here the normal distribution can be used as an approximation.

$$\text{Mean } \mu = np = 24 \times 0.20 = 4.8$$

$$\text{Variance } \sigma^2 = np(1-p) = 24 \times 0.20 \times 0.80 = 3.84$$

Standard deviation = 1.96

$$u = \frac{3.5 - 4.8}{1.96} = \frac{-1.3}{1.96} = -0.66$$

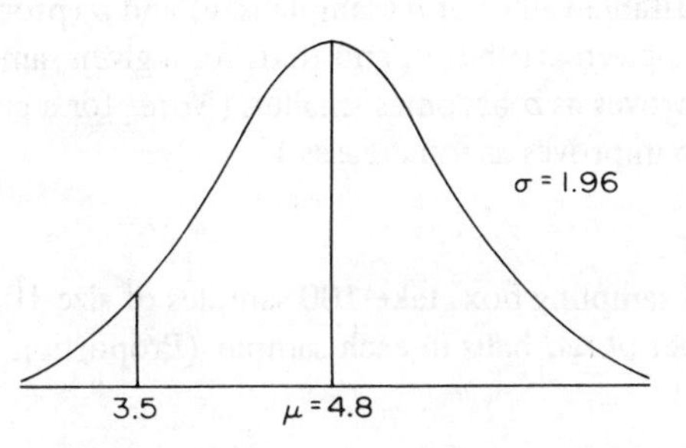

Figure 5.8

Table 3* gives the probability of three or less dustcarts being out of service as 0.2546

∴ Probability of more than three dustcarts being out of service

$$P(>3) = 1 - 0.2546 = 0.7454 \text{ or } 74.5\%$$

7. Here this is the hypergeometric distribution and since the sample size 4 is greater than 10% of population (20) no approximation can be made. Thus the hypergeometric distribution must be used.

Probability of 0 defects

$$P(0) = \frac{\binom{4}{0}\binom{16}{4}}{\binom{20}{4}} = \frac{\frac{16!}{12!\,4!}}{\frac{20!}{16!\,4!}} = \frac{16}{20} \times \frac{15}{19} \times \frac{14}{18} \times \frac{13}{17} = 0.3756$$

5.5 Practical Laboratory Experiment and Demonstrations

Using the binomial sampling box, experiment 8 in the laboratory manual demonstrates how the Poisson distribution can be used to approximate the binomial.

The laboratory instructions are given together with recording, analysis and summary sheets in pages 37-39 of the manual.

The laboratory instruction sheet for experiment 8 is reproduced in Appendix 1.

Appendix 1—Instruction Sheet for Experiment 8

Number of persons: 2 or 3.

Object

To demonstrate that the Poisson law may be used as an approximation to the binomial law for suitable values of n (sample size) and p (proportion of the population having a given attribute), and that, for a given sample size n, the approximation improves as p becomes smaller. (*Note:* for a given value of p, the approximation also improves as n increases.)

Method

Using the binomial sampling box, take 100 samples of size 10, recording, in table 21, the number of *red* balls in each sample. (Proportion of red balls in the population = 0.02.)

Analysis

1. Summarise the data into a frequency distribution of number of red balls per sample in table 22 and compare the experimental probability distribution with the theoretical binomial (given) and Poisson probability distributions.

Draw both the theoretical Poisson (mean = 0.2) and the experimental probability histograms on figure 1 below table 22.

2. Using the data of experiment 7 and table 23, compare the observed probability distribution with the binomial and Poisson (mean = 1.5) probability distributions.

Also, draw both the theoretical Poisson (mean = 1.5) and the experimental probability histograms on figure 2 below table 23.

Note: Use different colours for drawing the histograms in order that comparison may be made more easily.

6 Distribution of linear functions of variables

6.1 Syllabus Covered

Variance of linear combinations of variates; distribution of sample mean; central limit theorem.

6.2 Résumé of Theory and Basic Concepts

6.2.1 *Linear Combination of Variates*

Consider the following independent variates $x, y, z, \ldots$ with means $\bar{x}, \bar{y}, \bar{z}, \ldots$ and variances $\sigma_x^2, \sigma_y^2, \sigma_z^2, \ldots$

Let $w_r = ax_r + by_r + cz_r + \ldots$ where a, b, c are constants.

Then w is distributed with mean $\bar{w} = a\bar{x} + b\bar{y} + c\bar{z} + \ldots$ and variance

$$\sigma_w^2 = a^2\sigma_x^2 + b^2\sigma_y^2 + c^2\sigma_z^2 + \ldots$$

Special Case 1 – Variance of Sum of Two Variates

Here $a = +1$, $b = +1$ and $c = 0$ as for all other constants, then

$$w_r = x_r + y_r$$

$$\bar{w} = \bar{x} + \bar{y}$$

$$\sigma_w^2 = \sigma_x^2 + \sigma_y^2$$

or the variance of the sum of two independent variates is equal to the sum of their variances.

Special Case 2 – Variance of Difference of Two Variates

Here $a = +1$, $b = -1$ and all other constants $= 0$.

$$w_r = x_r - y_r$$

and

$$\bar{w} = \bar{x} - \bar{y}$$

$$\sigma_w^2 = \sigma_x^2 + \sigma_y^2$$

or the variance of the difference of two variates is the sum of their variances.

Note: It should be noted that while this theorem places no restraint on the form of distribution of variates the following conditions are of prime importance:

(1) If variates $x, y, z, \ldots$ are normally distributed then w is also normally distributed.

(2) If variates x, y, z are Poisson distributed then w is also distributed as Poisson.

Examples

1. In fitting a shaft into a bore of a housing, the shafts have a mean diameter of 50 mm and standard deviation of 0.12 mm. The bores have a mean diameter of 51 mm and standard deviation of 0.25 mm. What is the clearance of the fit?

The mean clearance = 51 – 50 = 1 mm
Variance of clearance = $0.12^2 + 0.25^2 = 0.0769$
Standard deviation of clearance = $\sqrt{0.0769} = 0.277$ mm

2. A machine producing spacers works to a nominal dimension of 5 mm and standard deviation of 0.25 mm. Five of these spacers are fitted on to a bolt manufactured to a nominal shaft dimension of 38 mm and standard deviation 0.50 mm.

What is the mean and variance of the clearance on the end of the shaft of the bolt?

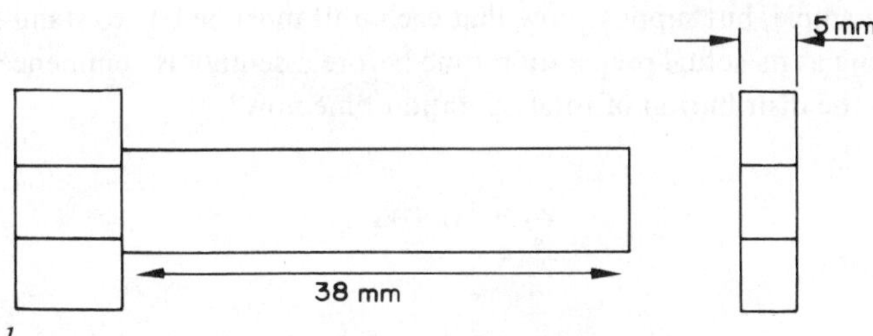

Figure 6.1

Here average clearance = 38 – 5 x 5 = 13 mm
Variance of clearance = $1 \times 0.5^2 + 5 \times 0.25^2 = 0.5625$ mm
Standard deviation = 0.75 mm

3. (*a*) The time taken to prepare a certain type of component before assembly is normally distributed with mean 4 min and standard deviation of 0.5 min. The time taken for its subsequent assembly to another component is independent of preparation time and again normally distributed with mean 9 min and standard deviation of 1.0 min.

What is the distribution of total preparation and assembly time and what proportion of assemblies will take longer than 15 min to prepare and assemble?

Let w = total preparation and assembly time for rth unit.

$$\bar{w} = 4 + 9 = 13 \text{ min}$$

$$\sigma_w^2 = 1^2 \times 0.5^2 + 1^2 \times 1.0^2 = 1.25$$

or standard deviation of w, $\sigma_w = \sqrt{1.25} = 1.12$ min

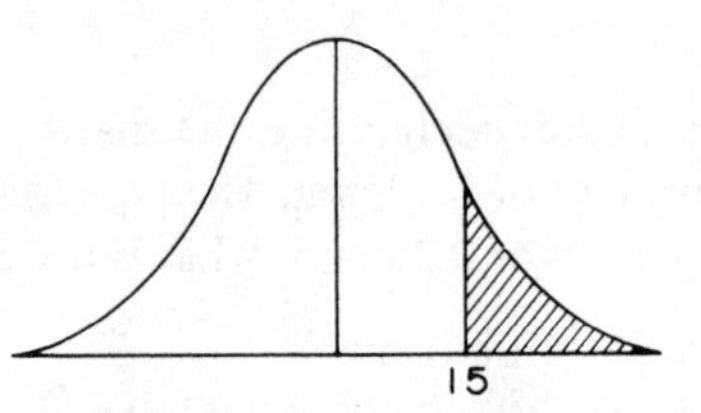

Figure 6.2

Distribution of preparation and assembly time, w

$$u = \frac{15-13}{1.12} = \frac{2}{1.12} = 1.78 \quad \text{mean} \quad \mu_w = 13 \qquad \sigma_w = 1.12$$

Reference to table 3* gives that probability of total assembly and preparation time exceeding 15 min is 0.0375 or 3.75% of units.

(*b*) In order to show clearly the use of constants $a, b, c, \ldots$, consider the previous example, but suppose now that each unit must be left to stand for twice as long as its actual preparation time before assembly is commenced.

What is the distribution of total operation time now?

Here

$$w_r = 3x_r + y_r$$

where

$$x_r = \text{preparation time}$$

$$y_r = \text{assembly time}$$

$$\bar{w} = (3 \times 4) + 9 = 21 \text{ min}$$

$$\sigma_w^2 = 3^2 \times 0.5^2 + 1^2 \times 1^2 = 3.25$$

Standard deviation of w = 1.8 min.

(*c*) To further clarify the use of constants, consider now example 3(*a*). Here the unit has to be sent back through the preparation phase twice before passing on to assembly.

Assuming that the individual preparation times are independent, what is the distribution of the total operation time now?

Here

$$w_r = (x_{r_1} + x_{r_2} + x_{r_3}) + y_r$$

or

$$\bar{w} = (4+4+4)+9 = 21 \text{ min as before}$$

however, variance

$$\sigma_w^2 = (1^2 \times 0.5^2 + 1^2 \times 0.5^2 + 1^2 \times 0.5^2) + 1^2 \times 1^2 = 1.75$$

Standard deviation

$$\sigma_w = 1.32$$

6.2.2 *Distribution of Sum of* n *Variates*

The sum of *n* equally distributed variates has a distribution whose average and variance are equal to *n* times the average and variance of the individual variates.

This follows direct from the general theorem in section 6.2.1.

Let

$$x = y = z \dots \quad \text{and} \quad a = b = c \dots = 1$$

then

$$\bar{w} = \bar{x} + \bar{x} + \dots + \bar{x} = n\bar{x}$$

$$\sigma_w^2 = (1^2 \times \sigma_x^2) + (1^2 \times \sigma_x^2) + \dots + (1^2 \times \sigma_x^2) = n\sigma_x^2$$

Example

Five resistors from a population whose mean resistance is 2.6 kΩ and standard deviation is 0.1 kΩ are connected in series. What is the mean and standard deviation of such random assemblies?

Average resistance = 5 x 2.6 = 13 kΩ
Variance of assembly = 5 x 0.1^2 = 0.05
Standard deviation = 0.225 kΩ

6.2.3 *Distribution of Sample Mean*

The distribution of means of samples of size *n* from a distribution with mean μ and variance σ^2 has a mean of μ and variance σ^2/n.

Population

mean μ
variance σ^2

Consider ith sample of size n from this population
Let x_r = rth member of this sample

then the mean of the ith sample

$$\bar{x}_i = \frac{1}{n}(x_1 + x_2 + x_3 + x_r + \ldots x_n)$$

$$= \left(\frac{1}{n}\right)x_1 + \left(\frac{1}{n}\right)x_2 + \left(\frac{1}{n}\right)x_3 + \ldots + \frac{1}{n}(x_r) + \ldots + \left(\frac{1}{n}\right)x_n$$

Since mean of x_1 = mean of x_2 = mean of $x_r = \mu$

average of distribution of samples of size $n = \frac{1}{n}(\mu + \mu + \ldots + \mu) = \mu$

Variance of distribution of sample of size $n = \left(\frac{1}{n}\right)^2 \sigma^2 + \left(\frac{1}{n}\right)^2 \sigma^2 + \ldots + \left(\frac{1}{n}\right)^2 \sigma^2$

$$= \frac{\sigma^2}{n}$$

Standard deviation of samples of size $n = \frac{\sigma}{\sqrt{n}}$

6.2.4 *Central Limit Theorem*

(Associated with theorem of distribution of sample mean in section 6.2.3 or with distribution of sum of variates in section 6.2.2.)

The distribution of sample mean (or the sum of n variates) has a distribution that is more normal than the distribution of individual variables.

This theorem explains the prominence of the normal distribution in the theory of statistics and the approximation to normality obviously depends upon the shape of the distribution of the variate and the size of n. As n increases the

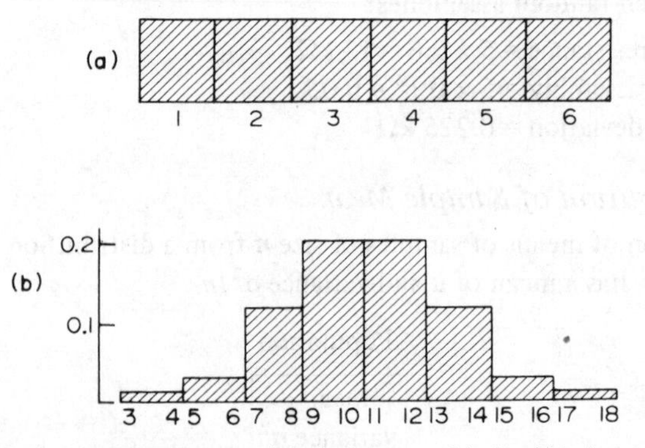

Figure 6.3. Probability distribution (a) the score of 1 die (b) the score of 3 dice.

sampling distribution of means gets closer to normality and similarly the closer the original distribution to normal the quicker the approach to true normal form.

However the rapidity of the approach is shown in figure 6.3 which shows the distribution of the total score of three 6-sided dice thrown 50 times. This is equivalent to sampling three times from a rectangular population and it will be seen that the distribution of the sum of the variates has already gone a long way towards normality.

6.2.5 *Distribution of the Sum (or Difference) of Two Means*

μ_x and μ_y are the means of distribution of x and y and σ_x^2, σ_y^2 their respective variances, then if a sample of n_x is taken from the x population and a sample of n_y from the y population, the distribution of the sum (or difference) between the averages of the samples has mean

$$\mu_x + \mu_y \ (\text{or } \mu_x - \mu_y)$$

and variance

$$\frac{\sigma_x^2}{n_x} + \frac{\sigma_y^2}{n_y}$$

In the special case where two samples of size n_1 and n_2 are taken from the same population with mean μ and variance σ^2, the moments of the distribution of sum (or difference) of sample averages is mean 2μ for sum (and 0 for difference) and variance

$$\sigma^2 \sqrt{\left(\frac{1}{n_1} + \frac{1}{n_2}\right)}$$

This theorem is most used for testing the difference between two populations, but this is left until chapter 7.

Example

A firm calculates each period the total value of sales orders received in £.p. The average value of an order received is approximately £400, and the average number of orders per period is 100.

What likely maximum error in estimation will be made if in totalling the orders, they are rounded off to the nearest pound?

Assuming that each fraction of £1 is equally likely (the problem can, however, be solved without this restriction) the probability distribution of the error on each order is rectangular as in figure 6.4, showing that each rounding off error is equally likely.

Consider the total error involved in adding up 100 orders each rounded off.

Statistically this is equivalent to finding the sum of a sample of 100 taken from the distribution in figure 6.4.

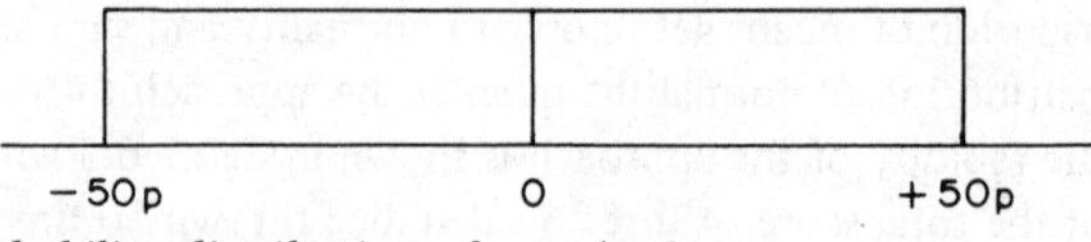

Figure 6.4. Probability distribution of error/order.

From theorem 3 the distribution of this sum will be normal and its mean and variance are given below.

Average error $= 0$
Variance of sum $= 100\sigma^2$ where σ^2 variance of the distribution of individual errors

For a rectangular distribution, $\sigma^2 = h^2/12$ where h = range of base.

$\therefore$ h in this problem = 100p = £1.00

$$\therefore \quad \text{Variance of sum} = \frac{100 \times 1}{12} = \frac{100}{12} = 8.33$$

Standard deviation = $\sqrt{8.33}$ = £2.90

Here the likely maximum error will be interpreted as the error which will be exceeded only once in 1000 times.

Since the error can be both positive or negative

$\therefore$ Maximum likely error = ± 3.29 x £2.9 or ±£9.50

6.3 Problems for Solution

1. A manufacturer markets butter in $\frac{1}{2}$ kg packages. His packing process has a standard deviation of 10 g. What must his process average be set at to ensure that the chance of any individual package's being 5% under the nominal weight of $\frac{1}{2}$ kg is only 5% (or 1 in 20)?

If the manufacturer now decides to market in super packages containing four $\frac{1}{2}$-kg packages, what proportion of his product can be saved by this marketing method if he still has to satisfy the condition that super packages must only have a 5% chance of being 5% under nominal weight of 2 kg?

2. The maximum payload of a light aircraft is 350 kg. If the weight of an adult is normally distributed (N.D.) with mean and standard deviation of 75 and 15 kg respectively, and the weight of a child is normally distributed with mean and standard deviation of 23 and 7 kg respectively, what is the probability that the plane can take off safely with

(*a*) four adult passengers?
(*b*) four adult passengers and one child?

In each case, what is the probability that the plane can take off if 40 kg of baggage is carried?

3. Two spacer pieces are placed on a bolt to take up some of the slack before a spring washer and nut are added. The bolt (b) is pushed through a plate (p) and then two spacers (s) added, as in figure 6.5.

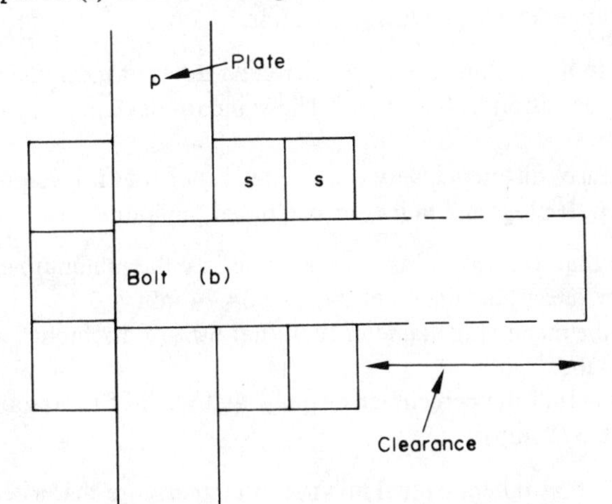

Figure 6.5

Given the following data on the production of the components

plate: mean thickness 12 mm, standard deviation of thickness 0.05 mm, normal distribution
bolt: mean length 25 m, standard deviation of length 0.025 mm, normal distribution
spacer: mean thickness 3 mm, standard deviation of thickness 0.05 mm, normal distribution

what is the probability of the clearance being less than 7.2 mm?

4. In a machine fitting caps to bottles, the force (torque) applied is distributed normally with mean 8 units and standard deviation 1.2 units. The breaking strength of the caps has a normal distribution with mean 12 units and standard deviation 1.6 units. What percentage of caps are likely to break on being fitted?

5. Four rods of nominal length 25 mm are placed end to end. If the standard deviation of each rod is 0.05 mm and they are normally distributed, find the 99% tolerance of the assembled rods.

6. The heights of the men in a certain country have a mean of 1.65 m and standard deviation of 76 mm.

(*a*) What proportion will be 1.80 m or over?

(*b*) How likely is it that a sample of 100 men will have a mean height as great as 1.68 m. If the sample does have a mean of 1.68 m, to what extent does it confirm or discredit the initial statement?

7. A bar is assembled in two parts, one 66 mm ± 0.3 mm and the other 44 mm ± 0.3 mm. These are the 99% tolerances. Assuming normal distributions, find the 99% tolerance of the assembled bar.

8. Plugs are to be machined to go into circular holes of mean diameter 35 mm and standard deviation of 0.010 mm. The standard deviation of plug diameter is 0.075 mm.

The clearance (difference between diameters) of the fit is required to be at least 0.05 mm. If plugs and holes are assembled randomly:

(*a*) Show that, for 95% of assemblies to satisfy the minimum clearance condition, the mean plug diameter must be 34.74 mm.

(*b*) Find the mean plug diameter such that 60% of assemblies will have the required clearance.

In each case find the percentage of plugs that would fit too loosely (clearance greater than 0.375 mm).

9. Tests show that the individual maximum temperature that a certain type of capacitor can stand is distributed normally with mean of 130°C and standard deviation of 3°C. These capacitors are incorporated into units (one capacitor per unit), each unit being subjected to a maximum temperature which is distributed normally with a mean of 118°C and standard deviation of 5°C.

What percentage of units will fail due to capacitor failure?

10. It is known that the area covered by 5 litres of a certain type of paint is normally distributed with a mean of 88 m^2 and a standard deviation of 3 m^2. An area of 3500 m^2 is to be painted and the painters are supplied with 40 5-litre tins of paint. Assuming that they do not adjust their application of paint according to the area still to be painted, find the probability that they will not have sufficient paint to complete the job.

11. A salesman has to make 15 calls a day. Including journey time, his time spent per customer is 30 min on average with a standard deviation of 6 min.

(*a*) If his working day is of 8 h, what is the chance that he will have to work overtime on any given day?

(*b*) In any 5-day week, between what limits is his 'free' time likely to be?

12. A van driver is allowed to work for a maximum of 10 h per day. His journey time per delivery is 30 min on average with a standard deviation of 8 min.

In order to ensure that he has only a small chance (1 in 1000) of exceeding the 10 h maximum, how many deliverties should he be scheduled for each day?

6.4 Solutions to Problems

1. At least 95% of individual packets must weigh more than 0.475 kg. Thus the process average weight must be set above 0.475 kg by 1.645 times the standard deviation (see figure 6.6; 5% of the tail of a normal distribution is cut off beyond 1.645 standard deviations), i.e at

$$0.475 + 1.645 \times 0.010 = 0.475 + 0.0164 = 0.491 \text{ kg}$$

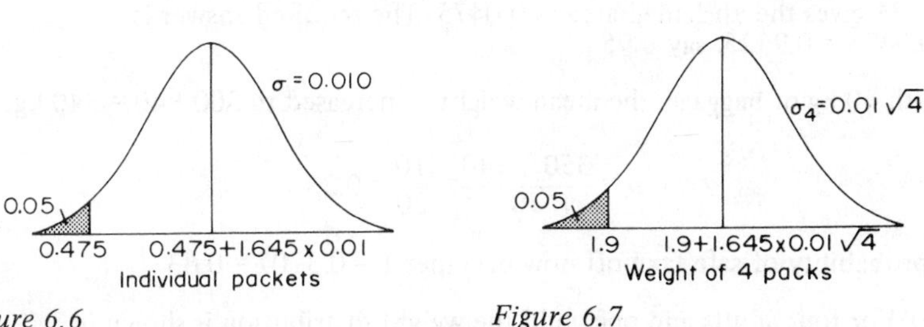

Figure 6.6 *Figure 6.7*

If individual packages are packed four at a time, the distribution of total net weight and the probability requirements are shown in figure 6.7.

The mean weight of 4 packages must be

$$1.9 + 1.645 \times 0.01\sqrt{4} = 1.9 + 0.033 = 1.933 \text{ kg}$$

Thus the process setting must be 1.933/4 = 0.483 kg

The long run proportional saving of butter per nominal $\frac{1}{2}$-kg package is

$$\frac{0.491 - 0.483}{0.491} = \frac{0.008}{0.491} = 0.0163 \text{ or } 1.63\%$$

2. (*a*) The weight of four adult passengers will be normally distributed with mean of 4 x 75(= 300) kg and standard deviation of $\sqrt{4} \times 15(=30)$ kg. The shaded area in figure 6.9 gives the probability that the plane is within its maximum payload.

The standardised normal variate,

$$u = \frac{350 - 300}{30} = \frac{50}{30} = 1.67$$

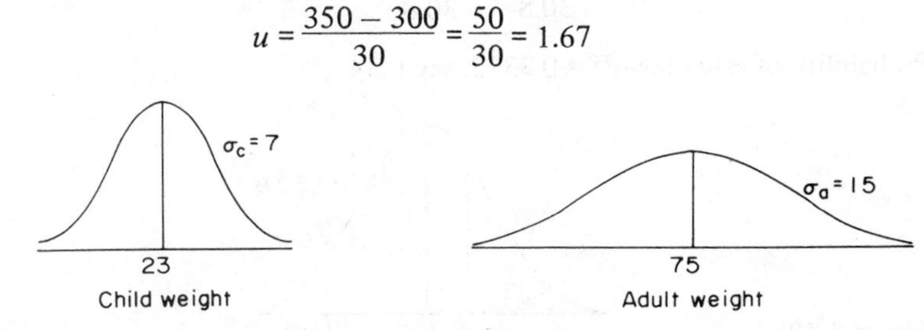

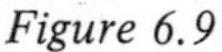

Figure 6.8 *Figure 6.9*

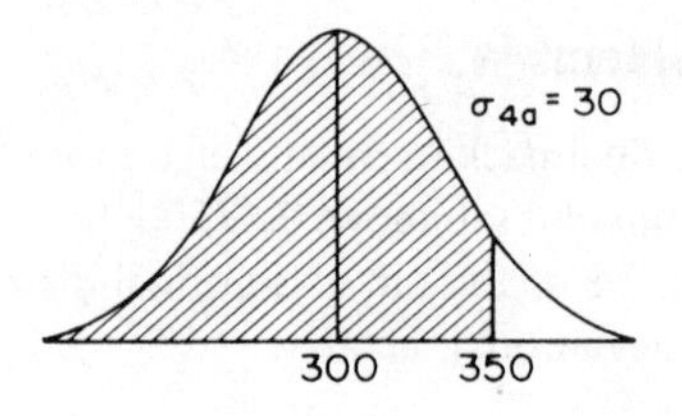

Figure 6.10

Table 3* gives the unshaded area as 0.0475. The required answer is $1 - 0.0475 = 0.9525$, say 0.95

With 40 kg of baggage, the mean weight is increased to $300 + 40 = 340$ kg.

$$u = \frac{350 - 340}{30} = \frac{10}{30} = 0.33$$

The probability of safe take-off now becomes $1 - 0.3707 = 0.63$

(*b*) For four adults and one child the weight distribution is shown in figure 6.11.

As before,

$$u = \frac{350 - 323}{30.8} = \frac{27}{30.8} = 0.88$$

Thus probability of safe take-off $= 1 - 0.1894 = 0.81$

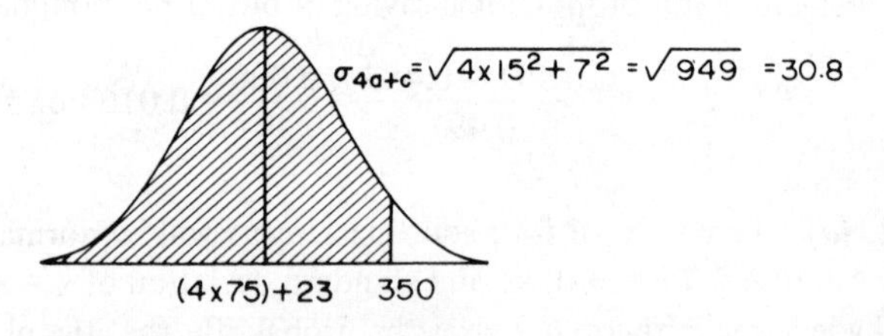

Figure 6.11

With 40 kg of baggage in addition, the distribution is

$$u = \frac{350 - 363}{30.8} = \frac{-13}{30.8} = -0.42$$

Probability of safe take-off $= 0.3372$, say 0.34

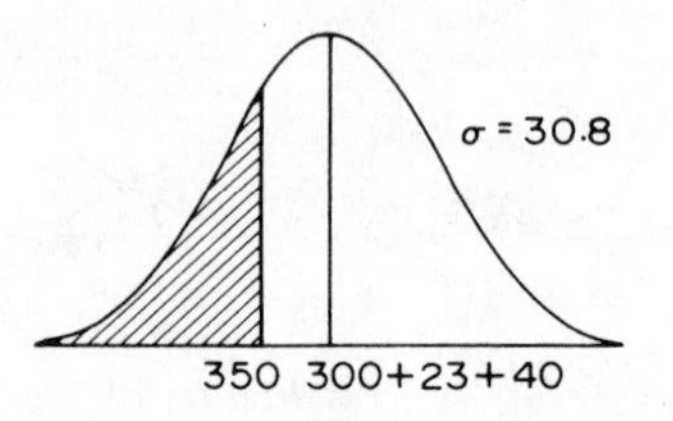

Figure 6.12

3. The mean clearance will be 25 – 12 – 3 – 3 = 7 mm

The variance of the clearance will be $0.025^2 + 0.050^2 + 0.050^2 + 0.050^2$

$$= 0.008125$$

and the standard deviation is 0.090 mm.

The distribution of clearance is shown in figure 6.13, the shaded area being the required answer.

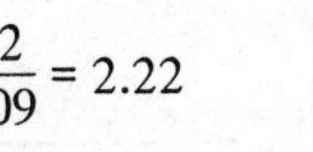

$$u = \frac{7.2 - 7.0}{0.09} = \frac{0.2}{0.09} = 2.22$$

thus the probability that the clearance is less than 7.2 mm is 1 – 0.0132 = 0.987

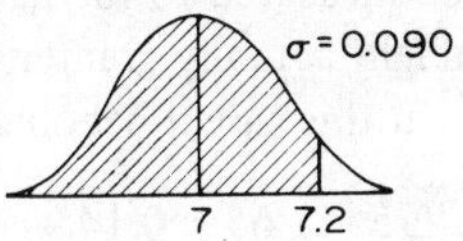

Figure 6.13

4. A cap will break if the applied force is greater than its breaking strength.

The mean excess of breaking strength is 12 – 8 = 4 units while the standard deviation of the excess of breaking strength is $\sqrt{(1.6^2 + 1.2^2)} = \sqrt{4.00} = 2.0$.

When the excess of cap strength is less than zero the cap will break and the proportion of caps doing so will be equal to the shaded area of figure 6.14, i.e the area below

$$u = \frac{0 - 4}{2} = -2 \quad \text{or} \quad 0.0228,$$

about $2\frac{1}{4}$% of caps.

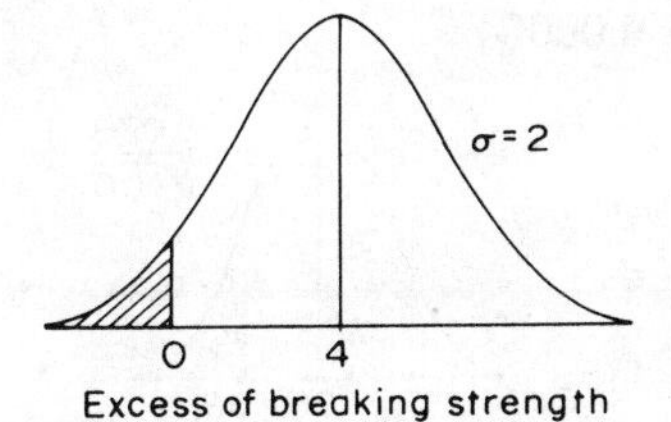

Figure 6.14

5. The distribution of the total length of four rods will be normal with a mean of 4 x 25 = 100 mm and standard deviation of $\sqrt{4} \times 0.05 = 0.10$ mm.

Ninety-nine per cent of all assemblies of four rods will have their overall length within the range

$$100 \pm 2.58 \times 0.10 \text{ mm} \quad \text{i.e.} \quad 100 \pm 0.26 \text{ mm}$$

6. (*a*) Assuming that heights can be measured to very small fractions of a

metre, the required answer is equal to the shaded area in figure 6.15.

$$u = \frac{1.80 - 1.65}{0.076} = 1.97$$

and area = 0.0244

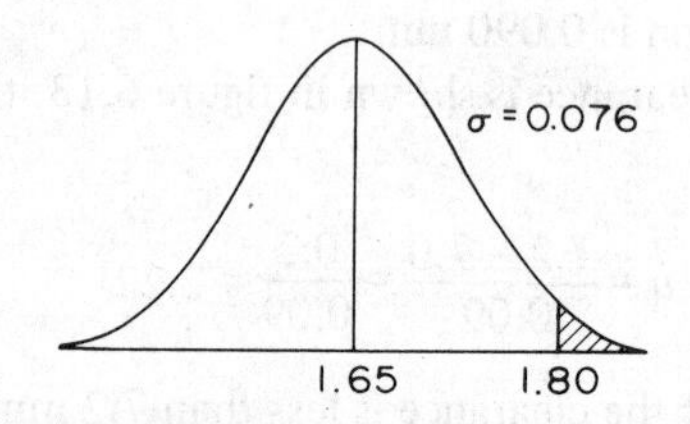

Figure 6.15

If heights, say, can only be measured to the nearest 5 mm, it would be reasonable to say that any height actually greater than 1.795 m would be recorded as 1.80 m or more. In this case u becomes

$$\frac{1.795 - 1.65}{0.076} = \frac{0.145}{0.076} = 1.91$$

Proportion over 1.80 m tall = 0.0281.

(*b*) Average heights of 100 men at a time (selected randomly) will be distributed normally with a mean of 1.65 m and standard deviation of $76/\sqrt{100} = 7.6$ mm.

Probability of a mean height of 1.68 m or more equals the shaded area in figure 6.16.

$$u = \frac{1.68 - 1.65}{0.0076} = 3.95$$

The shaded area is about 0.00004.

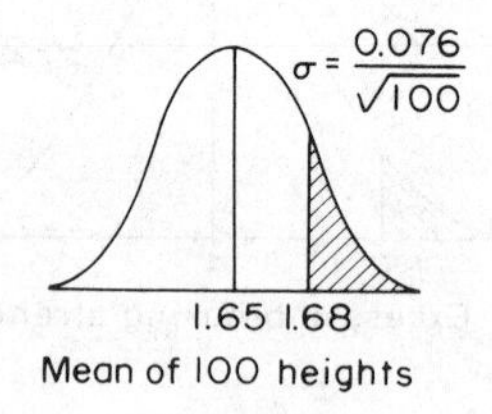

Figure 6.16

Possible alternative conclusions are that this particular sample is a very unusual one or that the assumed mean height of 1.65 m is wrong (being an underestimate) or that the standard deviation is actually higher than the assumed value of 76 mm.

7. The standard deviation of each component part is 0.3/2.58. The standard deviation of an assembly of each part will be $0.3\sqrt{2}/2.58$ about a mean of

66 + 44 = 110 mm.

Ninety-nine per cent of assemblies will lie within $110 \pm 0.3\sqrt{2}$ mm, i.e. within 110 ± 0.42 mm.

8. The distribution of clearance will have standard deviation of

$$\sqrt{(0.100^2 + 0.075^2)} = 0.125 \text{ mm}$$

(*a*) For 95% of assemblies to have clearance greater than 0.05 mm and assuming normality of the distribution, the mean must be

$$0.05 + 1.645 \times 0.125 = 0.256 \text{ mm}$$

and thus the mean plug diameter must be less than 35 mm by this amount, i.e. 34.74 mm.

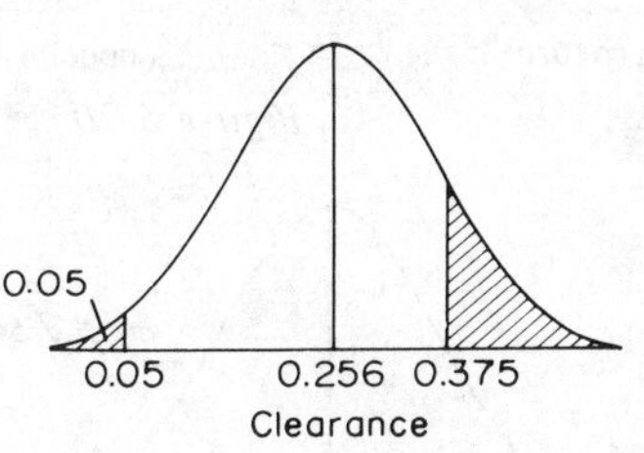

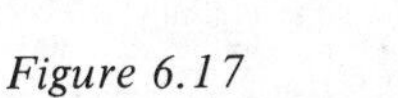

Figure 6.17

A clearance of 0.375 mm is equivalent to

$$u = \frac{0.375 - 0.256}{0.125} = 0.95$$

Table 3* shows that the probability of exceeding a standardised normal variate of 0.95 is 0.1711, i.e. approximately 17% of plugs would be too loose a fit.

(*b*) For 60% of assemblies to have clearance greater than 0.05 mm, the mean clearance must be

$$0.05 + 0.253 \times 0.125 = 0.082 \text{ mm}$$

and the mean plug diameter must be less than 35 mm by this amount, i.e. 34.92 mm.

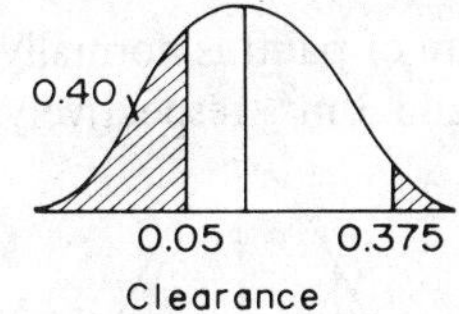

Figure 6.18

For a clearance of 0.375 mm,

$$u = \frac{0.375 - 0.082}{0.125} = 2.34$$

corresponding to an upper tail area of 0.0096, i.e. less than 1% of clearances would be too great.

9. A capacitor will fail if the maximum temperature to which it is subjected is greater than its own threshold temperature.

The proportion of units for which the maximum applied temperature is greater than the temperature the capacitor can resist is given by the shaded area in figure 6.21.

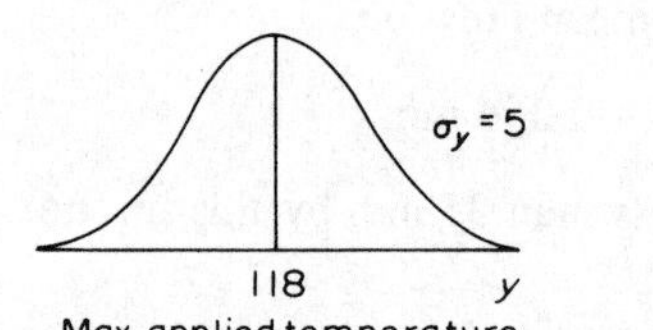

Figure 6.19

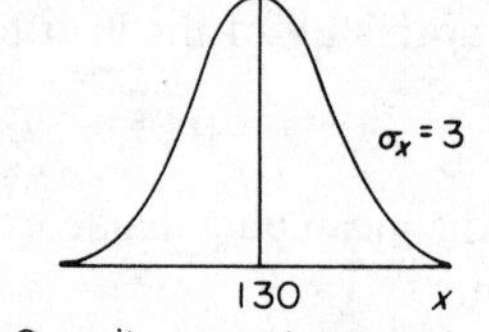

Figure 6.20

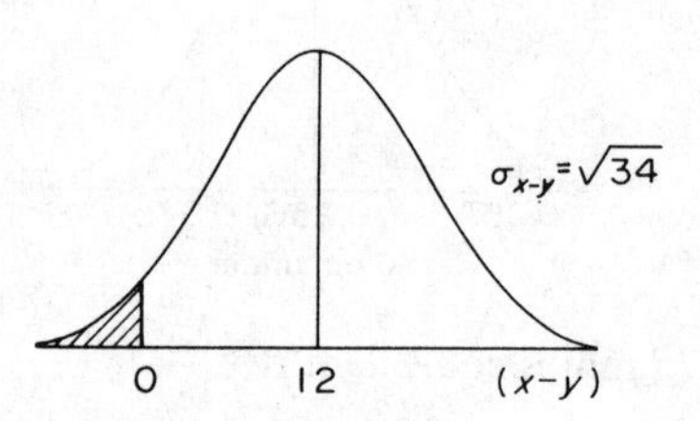

Figure 6.21

For the distribution of excess of capacitor threshold temperature over applied temperature,

the mean = 130 − 118 = 12 and variance = $3^2 + 5^2 = 34$.

Since this distribution will be normal and any negative excess corresponds to failure of the capacitor, we require the area below 0. The u-value equivalent to this is

$$u = \frac{0 - 12}{\sqrt{34}} = -2.06$$

giving a proportion of 0.0197.

10. The area covered by 5 litre of paint is normally distributed with mean and standard deviation of 88 m^2 and 3 m^2, respectively. Thus the area covered by

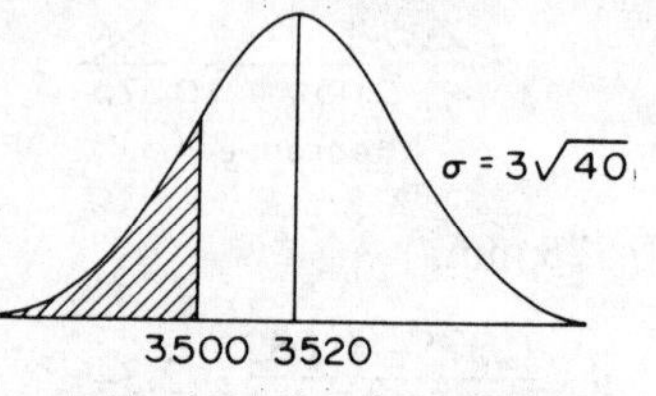

Figure 6.22 Area covered by 40 x 5 litres of paint

40 x 5 litre of paint will also be normally distributed with mean of 40 x 88 (=3520) m^2 and standard deviation of $\sqrt{40} \times 3$(=19.0) m^2.

The probability of covering less than 3500 m^2 is the probability of having insufficient paint for the job (shaded area of figure 6.22).

To find the shaded area

$$u = \frac{3500 - 3520}{19.0} = -1.05$$

giving an answer of about 14.7%.

11. (*a*) The distribution of time spent on a total of 15 calls will be approximately normal (by the central limit theorem) with a mean of 15 x 30(=450) min and a standard deviation of $\sqrt{15} \times 6$(=23.3) min.

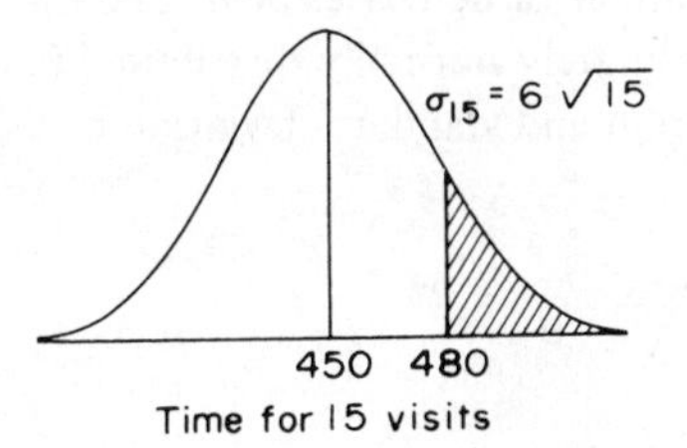

Figure 6.23

The probability that 15 calls take longer than 8 h is represented by the shaded area in figure 6.23.

480 min (8 h) corresponds to

$$u = \frac{480 - 450}{6\sqrt{15}} = 1.29$$

The required probability is 0.0985.

(*b*) There may be differing interpretations about what is meant by 'free' time in a week. 'Free' time for the salesman occurs on days when he works less than 8 h. The total of such time is found for five consecutive days, no account being taken of any 'overtime' that has to be worked. The solution of such a problem is quite difficult.

In this case, we shall consider 'free' time as the net amount by which his actual working time is less than his scheduled working time.

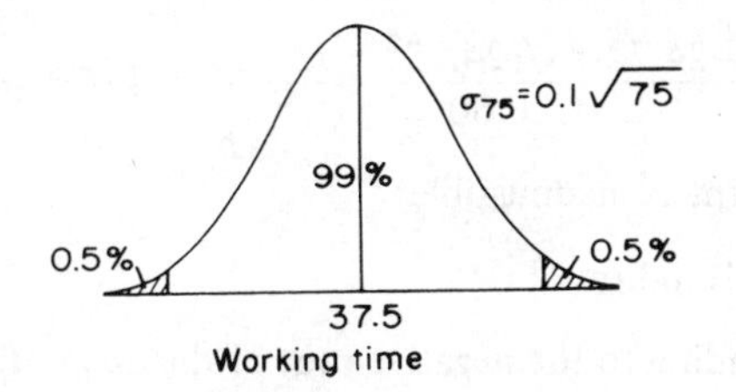

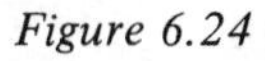
Figure 6.24

In 5 days, the number of calls to be made is 75. The distribution of total time to make these calls is approximately normal (by the central limit theorem) with a mean of 75 x 0.5 (= 37.5) h and standard deviation of $\sqrt{75} \times 0.1$ (= 0.866) h.

The salesman's total working time will lie within 2.58 standard deviations of the expected time for 75 calls with a probability of 99%, i.e. within

$$37.5 \pm 2.58 \times 0.1\sqrt{75} = 37.5 \pm 2.24 = 35.26 \text{ to } 39.74 \text{ h}$$

There is thus only a small chance (1%) that his 'free' time in one week lies outside the range

$$0.25 \text{ to } 4.75 \text{ h}$$

12. Let the required number of deliveries be n. The time required for n deliveries will be approximately normally distributed (central limit theorem) with mean time of $30n$ min and standard deviation of $8\sqrt{n}$ min.

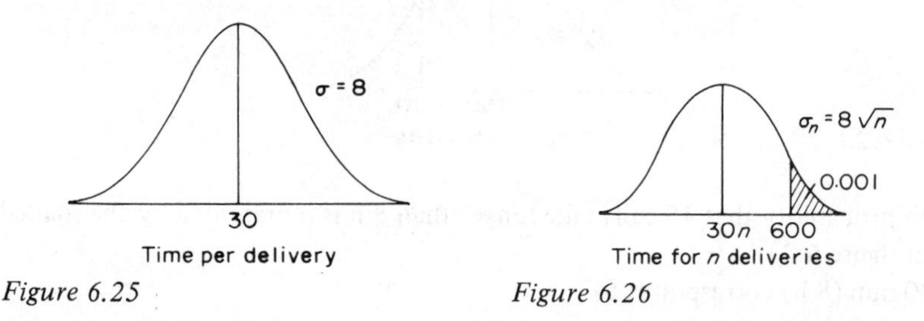

Figure 6.25 *Figure 6.26*

In order that there is only 1 chance in 1000 that n journeys take longer than 10 h (600 min), n must be such that

$$30n + 3.09 \times 8\sqrt{n} \leqslant 600$$

The largest value of n that satisfies the inequality can be found by systematic trial and error. However, a more general approach is to solve the equality as a quadratic in $\sqrt{n}$, taking the integral part of the admissible solution as the number of deliveries to be scheduled.

Thus

$$30n + 24.72\sqrt{n} - 600 = 0$$

$$\sqrt{n} = \frac{-24.72 \pm \sqrt{(24.72^2 + 72\,000)}}{60} = 4.0788 \text{ or } -4.9028$$

(discard negative term as inadmissible)

$$n = 16.64 \text{ or } 24.04$$

$n = 24.05$ corresponding to the negative root of the quadratic is clearly inadmissible since the average total journey time would be 12 h, violating the probability condition.

The number of deliveries to be scheduled is therefore 16.

If 16 deliveries were scheduled, the probability of exceeding 10 h would actually be less than 0.001–in fact about 1 in 10 000.

6.5 Practical Laboratory Experiments and Demonstrations

The following experiment from *Basic Statistics Laboratory Instruction Manual* demonstrates the basic concepts of the distribution of sample means and the central limit theorem.

Appendix 1–Experiment 12

Sampling Distribution of Means and Central Limit Theorem

Number of persons: 2 or 3.

Object

To demonstrate that the distribution of the means of samples of size n, taken from a rectangular population, with standard deviation σ tends towards the normal with standard deviation $\sigma/\sqrt{n}$.

Method

From the green rod population M6/3 (rectangularly distributed with mean of 6.0 standard deviation of 0.258), take 50 random samples each of size 4, replacing the rods after each sample and mixing them, before drawing the next sample of 4 rods.

Measure the lengths of the rods in the sample and record them in table 33.

Analysis

1. Calculate, to 3 places of decimals, the means of the 50 samples and summarise them into a grouped frequency distribution using table 34.

2. Also in table 34, calculate the mean and standard deviation of the sample means and record these estimates along with those of other groups in table 35.

Observe how they vary amongst themselves around the theoretically expected values.

3. In table 36, summarise the frequencies obtained by all groups and draw, on page 57, the frequency histogram for the combined results. Observe the shape of the histogram.

Sample no.	1	2	3	4	5	6	7	8	9	10
Total										
Average										

Sample no.	11	12	13	14	15	16	17	18	19	20
Total										
Average										

Sample no.	21	22	23	24	25	26	27	28	29	30
Total										
Average										

Sample no.	31	32	33	34	35	36	37	38	39	40
Total										
Average										

Sample no.	41	42	43	44	45	46	47	48	49	50
Total										
Average										

Table 6.1 (Table 33 of the laboratory manual)

Class Interval * units	Mid point	'Tally-marks'	Frequency f	Class u	fu	fu^2
5.600 – 5.650	5.625			–5		
5.675 – 5.725	5.700			–4		
5.750 – 5.800	5.775			–3		
5.825 – 5.875	5.850			–2		
5.900 – 5.950	5.925			–1		
5.975 – 6.025	6.000			0		
6.050 – 6.100	6.075			1		
6.125 – 6.175	6.150			2		
6.200 – 6.250	6.225			3		
6.275 – 6.325	6.300			4		
6.350 – 6.400	6.375			5		
		Totals of +ve terms				
		Total of –ve terms				
		Net totals				

Table 6.2 (Table 34 of the laboratory manual)

Calculation of Distribution Mean, x̄, *and Standard Deviations,* s

6.000 is the mid point of the class denoted by $u = 0$

Class width = 0.075

The mean, $\bar{x}$, of the distribution is given by

$$\bar{x} = 6.000 + 0.075\,\frac{\Sigma fu}{\Sigma f} = \qquad = \underline{\qquad\qquad}$$

The standard deviation, s', of the distribution is given by:

$$s' = 0.075\left[\frac{\Sigma fu^2 - \dfrac{(\Sigma fu)^2}{\Sigma f}}{\Sigma f}\right] = \qquad = \qquad = \underline{\qquad\qquad}$$

*** Strictly the class intervals should read 5.5875–5.6625 and the next 5.6625–5.7375 etc. but the present tabulation makes summarising simpler.**

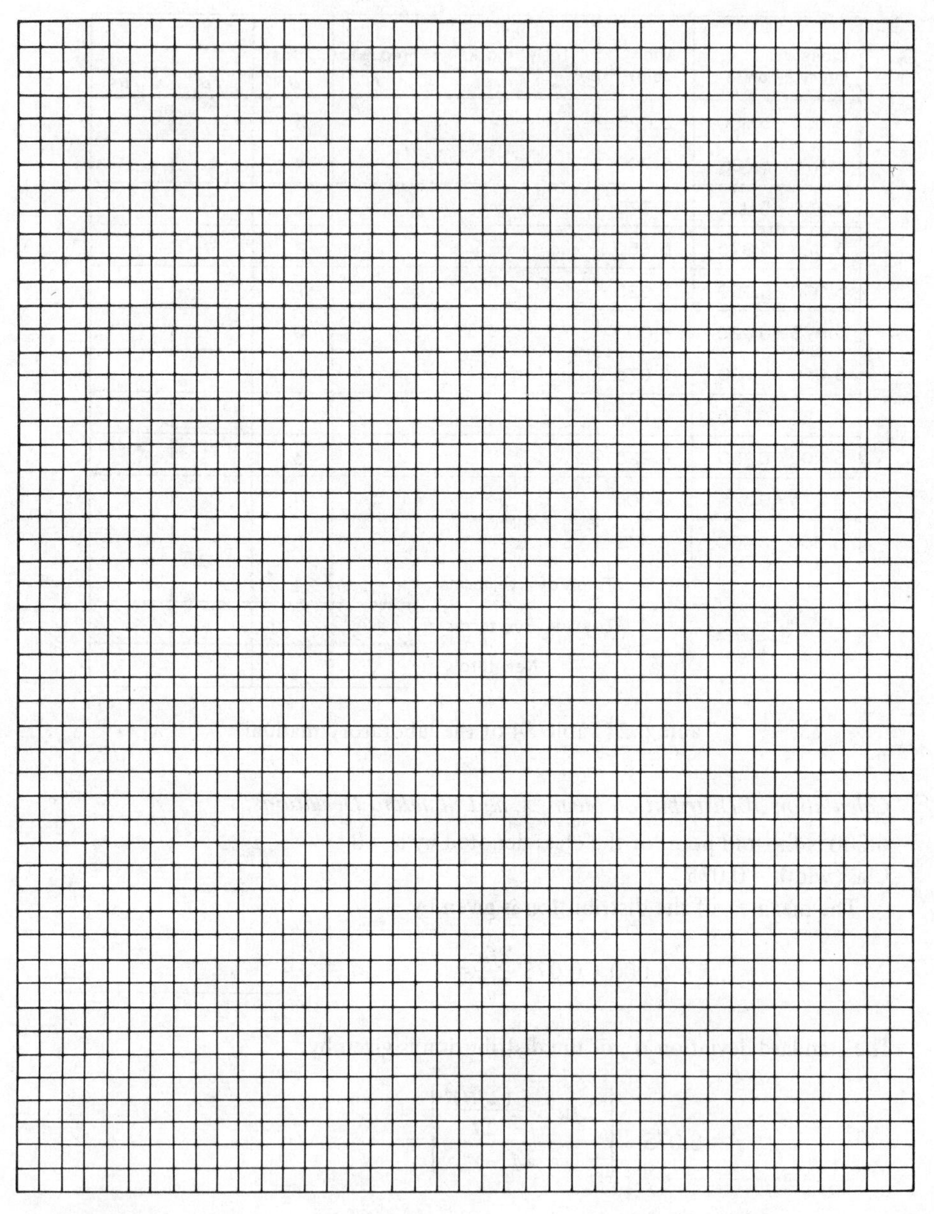

Figure 6.27. 'Page 57' of the laboratory manual.

7 Estimation and significance testing [I]—'large sample' methods

7.1 Syllabus

Point and interval estimates; hypothesis testing; risks in sampling; tests for means and proportions; sample sizes; practical significance; exact and approximate tests.

7.2 Résumé of Theory

7.2.1 *Point Estimators*

A point estimator is a number obtained from a sample and used to estimate a population parameter. For example, the average of a random sample is an estimator of the mean of the population from which it came. The sample median can also be used to estimate the mean of a symmetrical population as can other sample statistics. There are certain statistically desirable properties that point estimators should possess (unbiasedness, consistency, efficiency, sufficiency) and which make one estimator better than another for a particular purpose.

However, regardless of the estimator used, it is necessary to allow for uncertainty due to sampling variation, i.e. the numerical value obtained from the sample will not be exactly the same as the parameter value and an interval must be defined within which we can be reasonably confident that the parameter lies.

7.2.2 *Confidence Intervals*

Two numbers are calculated to determine the ends of an interval within which we can state that the population parameter lies. A probability is attached to the calculated interval and signifies the confidence we have in stating that the parameter actually falls within the interval. What this means is that if we found, say, a 95% confidence interval for a parameter, *if* such an interval were calculated for each of a large number of individual sample estimates, 95 out of every 100 intervals in the long run would contain the parameter and five would not.

The determination of confidence intervals is calculated from the sampling distribution of the particular sample estimator being used.

7.2.3 *Hypothesis Testing*

In science, a theory is developed to 'explain' the occurrence of an observed phenomenon. Further observations, usually coupled with deliberate experiments, are made to test the theory. The theory will be accepted as an adequate model until observations are made which it cannot satisfactorily 'explain'. In this case modification, or abandonment of the theory in favour of another one, is necessary. This is the approach used in statistical hypothesis testing.

An hypothesis is set up concerning a population; for example, it may be a statement about the value of one or more parameters of the population or perhaps about its form, i.e. that it is normal or exponential, etc. Statistical techniques are necessary to decide whether observations agree with such hypotheses because variation, and hence uncertainty, is usually present.

A statistical hypothesis is usually of the null type. As examples, consider the following.

(1) In testing whether a coin is biased, the hypothesis would be set up that it was fair, i.e. the probability of a 'head' on one toss is 0.5.
(2) In testing the efficiency of a new drug, it would be assumed as a hypothesis that it was no different in cure potential from the standard drug in current use.
(3) A new teaching method has been introduced; to assess whether it gives an improvement in its end product compared with the previous method, the hypothesis set up would be that it made no difference, i.e. that it was of the same effectiveness.
(4) To determine whether an overall 100 k.p.h. speed limit on previously unrestricted roads reduces accidents, the hypothesis would be set up that it makes no difference. The same method would be used to assess the effect of breathalyser tests.

7.2.4 *Errors Involved in Hypothesis Testing*

In deciding, on the basis of observed data subject to variation, whether or not to accept a statistical hypothesis, two types of error may be made.

Type I error (the α risk)

This first kind of error is the risk of rejecting the original hypothesis when it is, in fact, true. The risk is expressed in probability terms and is of magnitude α.

Type II error (the β risk)

This error is the risk of accepting (or better, failing to reject) the original hypothesis when, in fact, it is false. As for α, the risk is expressed as a probability of magnitude β.

7.2.5 *Hypothesis (Significance) Testing*

A test of a statistical hypothesis is a procedure for deciding whether or not to accept the hypothesis. This decision is made by assessing the significance of the observed results, i.e. are they so unlikely on the basis of the test hypothesis that the latter must be rejected in favour of some alternative hypothesis?

For example, in (1) in section 7.2.3 on testing the bias of a coin, the test would consist of counting the number of 'heads' in some convenient number of tosses and calculating the probability that such a result could have been obtained if the hypothesis was true.

When this probability has been calculated and it turns out to be very small, two explanations are possible. Either the hypothesis is false or else a rare event has occurred by chance.

It is customary to choose the first of these two alternatives when the probability is below a given level. In fact it will be seen that this probability is the risk of rejecting the hypothesis when in fact it is true. The levels of α are arbitrary but conventional values used are

$$\alpha = 0.05, \qquad \alpha = 0.01 \quad \text{and} \quad \alpha = 0.001$$

7.2.6 *Sample Size*

The magnitude of α can be fixed for a given test but β depends on the variability of the basic variable, the extent to which the test hypothesis is false (if it is false) and n, the sample size used. Generally the only one of these that can be altered at will is the sample size although there may be practical limitations of time, cost, or feasibility restricting even this.

Nevertheless, it is useful to know the sample size required to achieve given levels of α and β under given conditions of variability and parameter value of the population.

7.2.7 *Tests for Means and Proportions*

This section deals with some of the standard tests for population means and proportions. In both cases the following problems are considered.

I. Is the sample mean (or proportion) consistent with the value of the population mean assumed under the test hypothesis?

II. Do two different sample means (or proportions) indicate a significant difference in population means?

III. What is a confidence interval for the mean of a population?

IV. What is a confidence interval for the difference between the means of two populations?

Table 7.1 summarises the requirements for tackling these four problems.

The notation used is:

Sample size n (n_1 and n_2 for problems II and IV)

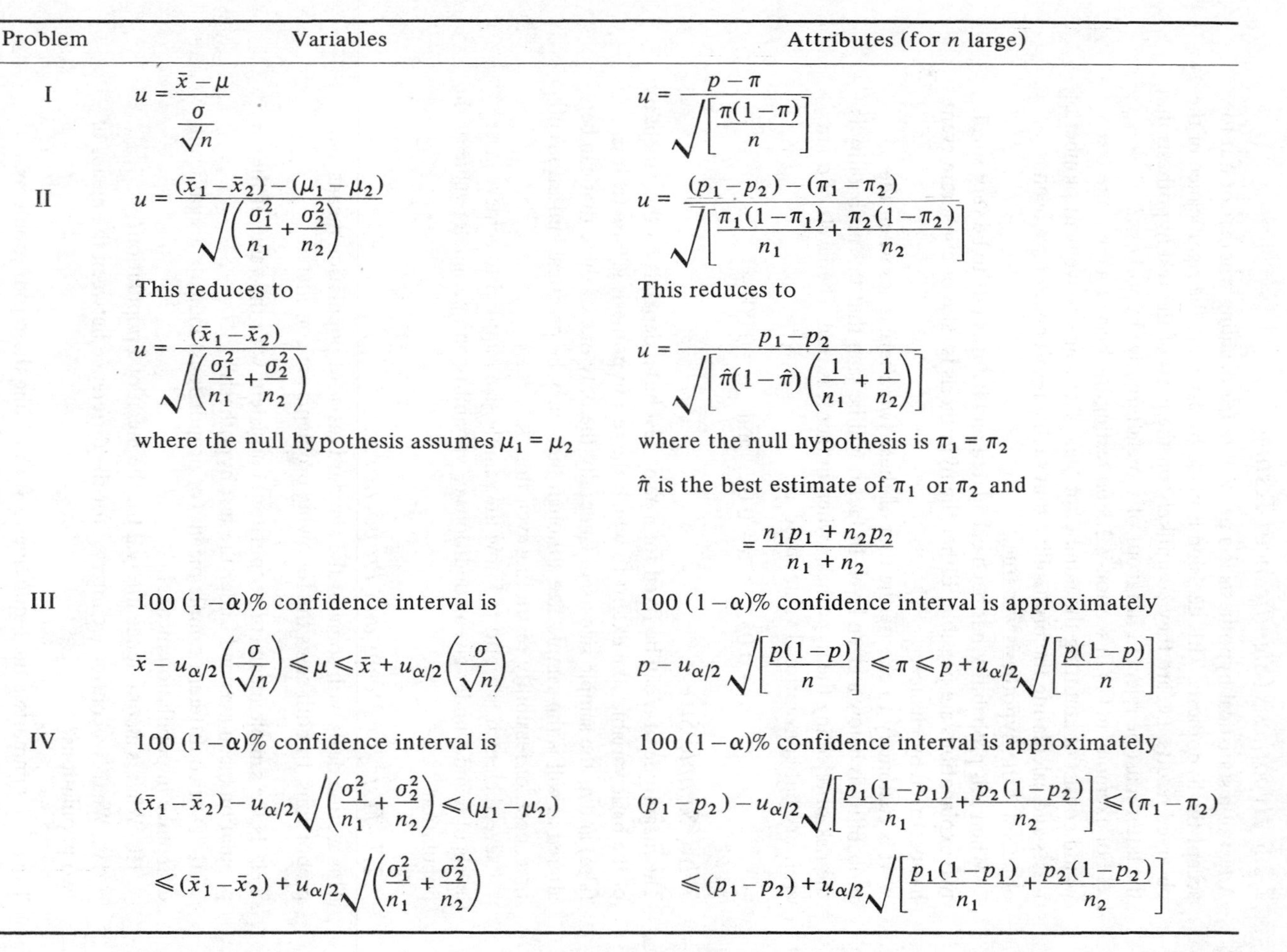

Problem	Variables	Attributes (for n large)
I	$u = \dfrac{\bar{x}-\mu}{\dfrac{\sigma}{\sqrt{n}}}$	$u = \dfrac{p-\pi}{\sqrt{\left[\dfrac{\pi(1-\pi)}{n}\right]}}$
II	$u = \dfrac{(\bar{x}_1-\bar{x}_2)-(\mu_1-\mu_2)}{\sqrt{\left(\dfrac{\sigma_1^2}{n_1}+\dfrac{\sigma_2^2}{n_2}\right)}}$ This reduces to $u = \dfrac{(\bar{x}_1-\bar{x}_2)}{\sqrt{\left(\dfrac{\sigma_1^2}{n_1}+\dfrac{\sigma_2^2}{n_2}\right)}}$ where the null hypothesis assumes $\mu_1 = \mu_2$	$u = \dfrac{(p_1-p_2)-(\pi_1-\pi_2)}{\sqrt{\left[\dfrac{\pi_1(1-\pi_1)}{n_1}+\dfrac{\pi_2(1-\pi_2)}{n_2}\right]}}$ This reduces to $u = \dfrac{p_1-p_2}{\sqrt{\left[\hat{\pi}(1-\hat{\pi})\left(\dfrac{1}{n_1}+\dfrac{1}{n_2}\right)\right]}}$ where the null hypothesis is $\pi_1 = \pi_2$; $\hat{\pi}$ is the best estimate of π_1 or π_2 and $= \dfrac{n_1p_1+n_2p_2}{n_1+n_2}$
III	$100\,(1-\alpha)\%$ confidence interval is $\bar{x}-u_{\alpha/2}\left(\dfrac{\sigma}{\sqrt{n}}\right) \leqslant \mu \leqslant \bar{x}+u_{\alpha/2}\left(\dfrac{\sigma}{\sqrt{n}}\right)$	$100\,(1-\alpha)\%$ confidence interval is approximately $p-u_{\alpha/2}\sqrt{\left[\dfrac{p(1-p)}{n}\right]} \leqslant \pi \leqslant p+u_{\alpha/2}\sqrt{\left[\dfrac{p(1-p)}{n}\right]}$
IV	$100\,(1-\alpha)\%$ confidence interval is $(\bar{x}_1-\bar{x}_2)-u_{\alpha/2}\sqrt{\left(\dfrac{\sigma_1^2}{n_1}+\dfrac{\sigma_2^2}{n_2}\right)} \leqslant (\mu_1-\mu_2) \leqslant (\bar{x}_1-\bar{x}_2)+u_{\alpha/2}\sqrt{\left(\dfrac{\sigma_1^2}{n_1}+\dfrac{\sigma_2^2}{n_2}\right)}$	$100\,(1-\alpha)\%$ confidence interval is approximately $(p_1-p_2)-u_{\alpha/2}\sqrt{\left[\dfrac{p_1(1-p_1)}{n_1}+\dfrac{p_2(1-p_2)}{n_2}\right]} \leqslant (\pi_1-\pi_2) \leqslant (p_1-p_2)+u_{\alpha/2}\sqrt{\left[\dfrac{p_1(1-p_1)}{n_1}+\dfrac{p_2(1-p_2)}{n_2}\right]}$

Table 7.1

Variables Population mean μ (μ_1 and μ_2 for problems II and IV)
Population standard deviation σ (σ_1 and σ_2 for problems II and IV) assumed known
Sample mean $\bar{x}$ ($\bar{x}_1$ and $\bar{x}_2$ for II and IV)

Proportions Population proportion (π_1 and π_2 for problems II and IV)
Sample proportion p (p_1 and p_2 for problems II and IV)

u is the standardised normal variate. In the case of variables, σ (and σ_1 and σ_2 as appropriate) is assumed to be known or calculated from a sample of size n larger than about 30.

7.2.8 *Practical Significance*

Even if a significant result is obtained, i.e. an observed sample is so extreme that the test hypothesis is no longer tenable, such *statistical* significance need not mean that the result is of any *practical* significance. For example, if a particular kind of light bulb has a mean life of 1220 h and a test of significance detects that, after a fairly expensive modification of the lamp-making process the mean life is increased, the modification is unlikely to be worth permanent incorporation if the new mean life is only 1225 h, say.

In summary then, the decision made as a result of a significance test depends on the possible consequences of that decision together with any other relevant information that may be available.

7.2.9 *Exact and Approximate Tests*

This section was given the sub-heading of 'Large Sample' Methods which is a common classification in the literature. Parts of chapter 8 refer to 'small sample' methods. The authors believe that a better classification would be into 'exact' and 'approximate' tests.

For example, in table 7.1, both of the tests using variables as well as the confidence interval estimation are exact for any n, provided that the populations involved are normal and that the variance σ^2 (and σ_1^2 and σ_2^2 as appropriate) is *known.* The tests and intervals become approximate when the populations involved are not normal but because of the central limit theorem, the error involved is very small for n larger than about 4.

Approximations are also introduced when σ is unknown but is estimated from the sample data. In this case, provided n is greater than about 30, very little error is involved. It is in cases where a test statistic is approximately normally distributed (which is often the case for large n) that the description of large sample methods is applied. Note, however, that the u-test can be exact for any n under appropriate conditions.

For attributes, as mentioned in table 7.1, all the procedures are approximate since they depend on the tendency of the binomial distribution towards normality for large n (and preferably with π neither small nor large)—see chapter 5.

7.2.10 *Interpretation of Significant Results*

Great care must be taken in the interpretation of a significant result. If a sample result is extreme (i.e. significantly different from expectation) the rules given lead to rejection of the test hypothesis. However, it is possible to get such a result when the hypothesis is true because the sample is not random but is heavily biased. This bias may arise either in the initial selection of the sample or in the subsequent extraction of numerical data from the members of the sample or in other subtle ways.

For example, in a coin-tossing experiment, the null hypothesis of a 50% probability of heads may be rejected by the sample evidence for a particular coin, the conclusion being that the coin is biased. However, the true situation could be that the coin is not biased but the method of tossing it (i.e. of sampling) is biased. This possibility should be considered in the initial design of the experiment so that such a mistake is not made in the final conclusion.

To repeat, there is more to statistics than knowing which formula to substitute the numbers into—the relevance and validity of the numbers must be considered and any interpretation of results closely matched to the circumstances of the case.

7.2.11 *Worked Examples*

1. From long experience, a variable is known to be normally distributed with standard deviation 6.0 about any given value of the mean, i.e. whatever the current mean is, the variability about that mean is constant. A random sample of 16 items from the population has a mean of 53.0. Is the current population mean 50.0?

Set up the test hypothesis H_0: mean (i.e. $E[x]$) = 50 and the alternative hypothesis $H_1 : E[x] \neq 50$.

The means of samples of size 16 will have standard error of $6/\sqrt{16} = 1.5$ and thus the sample mean of 53 deviates from the overall assumed mean by

$$u = \frac{53-50}{1.5} = 2 \text{ standard errors}$$

Since this is a two-sided test (i.e. if the mean is not 50, there is no knowledge that it *must* be larger), the result is significant at the 5% level as the observed value of $|u|$ is greater than 1.96. In fact, the actual significance level corresponds to about 4.5%. If the consequences of wrong rejection of the hypothesis were not very serious—as measured in terms of money, safety, inconvenience—then it would be reasonable to reject the assumption that the mean of the population is currently 50 units.

In general, having shown that there is some evidence (but not complete proof) that the mean is not 50 units, the next quesiton is—what is it?

The sample mean provides the best estimate of the population mean but an allowance must be made for sampling fluctuations; this is done by using the standard error of the sample mean to determine a confidence interval for the population mean.

For 95% confidence, the interval (conventionally symmetric in tail probability) is

$$\bar{x} - 1.96\sigma/\sqrt{n} \quad \text{and} \quad \bar{x} + 1.96\sigma/\sqrt{n} \quad \text{i.e.}$$

$$53 - 1.96 \times 1.5 \quad \text{and} \quad 53 + 1.96 \times 1.5$$

$$53 - 2.94 \quad \text{and} \quad 53 + 2.94$$

$$50.06 \quad \text{and} \quad 55.94 \quad \text{say} \quad 50.1 \quad \text{and} \quad 55.9$$

Notice that the interval does not include the previously assumed mean of 50.0. In this respect, the two procedures (hypothesis testing and interval estimation) are equivalent since the test hypothesis will be rejected at the 5% level of significance if the observed sample mean is more than 1.96 standard errors on either side of the assumed mean, and if this is the case the 95% confidence interval cannot include the assumed mean. This argument applies in the two-sided case for any significance level α and associated confidence probability $(1 - \alpha)$.

Also note, that in this example, the standard deviation was known and the test and confidence interval estimation was perfectly valid for any size of sample.

2. A synthesis of pre-determined motion times gives a nominal time of 0.10 min for the operation of piecing-up on a ring frame, compiled after analysis of the standard method. 160 observed readings of the piecing-up element had an average of 0.103 min and standard deviation of 0.009 min. Is the observed time really different from the nominal time?

Here, the population σ is not known but an estimate based on 160 (random) readings will be satisfactory.

Set up H_0: real mean element time, μ_0, = 0.100 min, H_1: real mean element time, μ_1, $\neq$ 0.100 min

$$u = \frac{\bar{x} - \mu_0}{\sigma/\sqrt{n}} \simeq \frac{0.103 - 0.100}{0.009/\sqrt{160}} = \frac{\sqrt{160}}{3} = 4.22$$

This is significant at the 1 in 1000 level ($|u| > 3.09$), the actual type I error being less than 6 parts in 100 000 (table 3*).

Ninety-nine per cent confidence limits for the real mean piecing-up time under the conditions applying during the sampling of the 160 readings are

$$0.103 \pm 2.58 \times \frac{0.009}{\sqrt{160}} = 0.103 \pm 0.00184, \quad \text{i.e.}$$

$$0.1012 \text{ to } 0.1048 \text{ min}$$

Thus, the evidence suggests that the synthesis of the mean operation time tends to underestimate the actual time by something between 1% and 5%. Whether this is of any practical importance depends on what use is going to be made of the synthetic time. Perhaps the method of synthesising the time may be worth review in order to bring it into line with reality.

3. In special trials carried out on two furnaces each given a different basic mix, furnace A in 200 trials gave an average output time of 7.10 h while 100 trials with furnace B gave an average output time of 7.15 h.

Given that, from previous records, the variance of furnace A is 0.09 h² and of B is 0.07 h² and an assurance that these variances did not change during the trials, is furnace A more efficient than B?

First of all, set up the test hypothesis that there is no difference in furnace efficiencies (i.e. average output times). The test is two-sided since there is no reason to suppose that if one is more efficient then it is known which one it will be.

Set up

$$H_0: \mu_A = \mu_B, \quad \text{i.e.} \quad \mu_A - \mu_B = 0$$

$$H_1: \mu_A - \mu_B \neq 0$$

The test statistic appropriate is

$$u = \frac{(\bar{x}_A - \bar{x}_2) - (\mu_A - \mu_B)}{\sqrt{\left(\dfrac{\sigma_A^2}{n_A} + \dfrac{\sigma_B^2}{n_B}\right)}}$$

which becomes on substituting the observed data and the test assumptions regarding $(\mu_A - \mu_B)$

$$u = \frac{(7.10 - 7.15) - 0}{\sqrt{\left(\dfrac{0.09}{200} + \dfrac{0.07}{100}\right)}} = \frac{-0.05}{0.034} = -1.47$$

Since this is numerically less than 1.96, or any higher value of u corresponding to a smaller α, the difference in mean output times has not been shown to be statistically significant at any reasonable type I error level.

Note: Even if a very highly significant value of u had been obtained (say $|u| > 4.0$) then the question could still not have been answered because of the way the trials had been set up. The two furnaces may have been different in mean output times (efficiencies) but because different basic mixes had always been used in the furnaces, it is not apparent how much of the efficiency difference was due to the different mixes and how much was due to inherent properties of the furnaces (including, perhaps, the crews who operate them). To determine whether the mix differences, furnace differences or a combination of both are responsible for differing mean output times would require a properly designed experiment (this experiment is not designed to answer the question posed).

In addition, it was assumed that the variances of the output times would be unchanged during the special trials. This may often be a questionable assumption and is unnecessary in this example since the sample variances of the 200 and 100 trials respectively could be substituted for σ_A^2 and σ_B^2 with very little effect on the significance test.

4. In a given year for a random sample of 1000 farms with approximately the same area given to wheat, the average yield of wheat per hectare (10 000 m^2) was 2000 kg, with standard deviation of 192 kg/ha. The following year, for a random sample of 2000 farms, the average was 2020 kg/ha, with standard deviation of 224 kg/ha. Does the second year show an increased yield?

In this case, because of the large samples, each of them greater than about 30, the sample variances can be used instead of the unknown population variances.

Set-up H_0: no difference in mean yield per hectare, i.e. $\mu_1 - \mu_2 = 0$
H_1: mean yields per hectare different between the years, i.e. $\mu_1 - \mu_2 \neq 0$

$$u \simeq \frac{(\bar{x}_1 - \bar{x}_2) - (\mu_1 - \mu_2)}{\sqrt{\left(\frac{\sigma_1^2}{n_1} + \frac{\sigma_2^2}{n_2}\right)}} = \frac{(2000 - 2020) - 0}{\sqrt{\left(\frac{192^2}{1000} + \frac{224^2}{2000}\right)}} = \frac{-20}{\sqrt{(36.9 + 25.1)}} = \frac{-20}{\sqrt{62}} = -2.54$$

This is almost significant at the 1% level and suggests that the mean yield for the whole population of farms is greater in the second year.

As a word of warning, such a conclusion may not really be valid since the two samples may not cover in the same way the whole range of climatic conditions, soil fertility, farming methods, etc. The significant result may be due as much to the samples' being non-representative as to a real change in mean yield for the whole population. The extent of each would be impossible to determine without proper design of the survey. There are many methods of overcoming this, one of which would be to choose a representative samples of farms and use the same farms in both years.

5. A further test of the types illustrated in examples 3 and 4 can be made when the population variances are unknown but there is a strong *a priori* suggestion that they are equal. In this case, the two sample variances can be pooled to make the test more efficient, i.e. to reduce β for given α and total sample size $(n_1 + n_2)$.

A group of boys and girls were given an intelligence test by a personnel officer. The mean scores, standard deviations and the numbers of each sex are given in table 7.2. Were the boys significantly more intelligent than the girls?

	Boys	Girls
Mean score	124	121
Standard deviation	11	10
Number	72	50

Table 7.2

The question as stated is trivial. If the test really does measure that which is termed 'intelligence', then on average *that* group of boys was more intelligent than *that* group of girls, although as a group they were more variable than the girls.

However, if the boys are a random sample from some defined population of boys and similarly for the girls, then any difference in average intelligence between the populations may be tested for.

Assuming that there is a valid reason for saying that the two populations have the same variances, the two sample variances can be pooled by taking a weighted average using the sample sizes as weights (strictly the degrees of freedom—see chapter 8—are used as weights but this depends on whether the degrees of freedom were used in calculating the quoted standard deviations; in any case, since the sample sizes here are large the error introduced will be negligible).

Pooled estimate of variance of individual scores

$$= \frac{(72 \times 11^2) + (50 \times 10^2)}{72 + 50} = 112$$

(*Note:* The variances are pooled, *not* the standard deviations.)

$$u = \frac{(\bar{x}_B - \bar{x}_G) - (\mu_B - \mu_G)}{\sqrt{\left(\frac{s^2}{n_B} + \frac{s^2}{n_G}\right)}}$$

where s^2 is the pooled variance

$$= \frac{(124 - 121) - 0}{\sqrt{[112(\frac{1}{72} + \frac{1}{50})]}} = \frac{3\sqrt{3600}}{\sqrt{(112 \times 122)}} = \frac{3 \times 60}{117} = 1.54$$

Thus there is no evidence that the populations of boys and girls differ in average intelligence. This conclusion does not mean that there is *not* a difference, merely that if there is one we have not sufficient evidence to demonstrate it, and even if we did, it may be so small as to be of no practical importance at all.

Confidence limits for the difference between two population means can be set in the same way as in examples (1) and (2) above.

Thus 95% confidence limits for the difference in mean intelligence are given by

$$(\bar{x}_B - \bar{x}_G) \pm 1.96\sqrt{\left(\frac{s_B^2}{n_B} + \frac{s_G^2}{n_B}\right)}$$

or, using the pooled variance, by

$$\bar{x}_B - \bar{x}_G \pm 1.96\sqrt{\left[s^2\left(\frac{1}{n_B} + \frac{1}{n_G}\right)\right]} = (124 - 121) \pm 1.96\sqrt{[112(\tfrac{1}{72} + \tfrac{1}{50})}$$

$$= 3 \pm 1.96\sqrt{\left(\frac{112 \times 122}{3600}\right)}$$

$$= 3 \pm 1.96 \times \tfrac{117}{60} = 3 \pm 3.82, \quad \text{i.e}$$

$$-0.82 \text{ to } 6.82$$

including the null value, 0, as it must from the significance test.

Note: The use of 1.96 instead of 2.0 is somewhat pedantic in practical terms; it is retained in this chapter to serve as a reminder that the appropriate u-factor is found from the tables* of the normal distribution in conjunction with the choice of α.

Examples Concerning Proportions

6. A programmed learning course has been introduced to train operators for a precision job in a company. Ten per cent of the operators trained by the previous method were found to be unsuitable for the job. Of 100 operators trained by the new method, eight were not suitable. Is the new method better than the old?

Set up the null hypothesis that both methods are the same in their effect and hence have the same 'failure' rate, i.e.

$$H_0: \pi = 0.10 \qquad H_1: \pi \neq 0.10$$

π is the probability that any one operator will not benefit from the course and is assumed constant for all operators.

The sample proportion of operators, p, who do not benefit from the course will be binomially distributed with mean of π and standard error of $\sqrt{[\pi(1-\pi)/n]}$.

For large n, this binomial distribution can be approximated by the normal distribution with the same parameters.

Thus,

$$u \triangleq \frac{p - \pi}{\sqrt{\left[\frac{\pi(1-\pi)}{n}\right]}} = \frac{0.08 - 0.10}{\sqrt{\left(\frac{0.1 \times 0.9}{100}\right)}} = -\frac{0.02}{0.03} = -0.67$$

Since this is not at all a low probability result, there is no evidence that the new method is any more or less effective than the previous one.

7. A certain type of seed is supposed to have a germination rate of 80%. If 50 seeds are tested and 14 fail to germinate, does this mean that the batch from which they came is below specification?

Set up

$$H_0: \pi = 0.80 \qquad H_1: \pi \neq 0.80$$

Use the normal approximation to give

$$u \simeq \frac{p-\pi}{\sqrt{\left[\frac{\pi(1-\pi)}{n}\right]}} = \frac{0.72-0.80}{\sqrt{\left(\frac{0.8 \times 0.2}{50}\right)}} = \frac{-0.08\sqrt{50}}{\sqrt{0.16}} = -1.41$$

This is not numerically large enough to reject the test hypothesis—the type I error would correspond to just under 16%.

A slight improvement can be made in the adequacy of the normal approximation by making the so-called correction for continuity. However, with the large sample sizes generally required for use of the normality condition, this refinement will not usually be worth incorporating. It is given here as an example.

36 or fewer germinating seeds can be considered as 36.5 or fewer on a continuous scale. 36.5 corresponds to 0.73 as a proportion of 50 and the corrected value for u becomes

$$u \simeq \frac{0.73-0.80}{\sqrt{\left(\frac{0.8 \times 0.2}{50}\right)}} = \frac{-0.07\sqrt{50}}{0.4} = -1.24$$

The type I error corresponding to such a value of u (two tails) is about 21.5%; too high for most people to contemplate making.

Both this example and example (6) could have been done using the *number* of occurrences rather than the *proportion* of occurrences in a sample. The approaches are identical but for setting confidence limits, the proportion method is better.

Standardising the number of 'successes' x in n trials gives

$$u \simeq \frac{x-n\pi}{\sqrt{[n\pi(1-\pi)]}}$$

which on dividing top and bottom by n gives

$$u \simeq \frac{\frac{x}{n} - \pi}{\sqrt{\left[\frac{n\pi(1-\pi)}{n^2}\right]}} = \frac{p-\pi}{\sqrt{\left[\frac{\pi(1-\pi)}{n}\right]}}$$

The exact test can be carried for this example since the appropriate parameters are tabulated in table 1* (cumulative binomial probabilities).

For an assumed germination rate of 80%, the expected (mean) number of seeds germinating out of 50 tested is 50 x 0.8 = 40. Because of the method of tabulating (i.e. for proportions $\leqslant 0.50$), the problem is best discussed in terms of seeds failing to germinate, the expected number being 50 x 0.2 = 10.

The probability of 14 or more failing to germinate is 0.1106 and the probability of six or fewer failing to germinate is 1–0.8966 = 0.1034, i.e. a total probability (magnitude of type I error) of 21.40% which compares favourably with the refined normal approximation.

A a further point, if a 5% significance level is specified for this problem (two-sided test since the true germination rate could be above or below 80%), using table 1* with $n = 50$ and π (tabulated as p) = 0.20, the acceptance region for failed seeds is from 5 up to 15 inclusive with the critical region split as near equally as possible between the two tails (1.85% in the lower tail and 3.08% in the upper tail).

Approximate confidence limits for the seed population germination rate are found as

$$95\% \quad p \pm 1.96\sqrt{\left[\frac{p(1-p)}{n}\right]} = 0.72 \pm 1.96\sqrt{\left(\frac{0.72 \times 0.28}{50}\right)} = 59.6\% \text{ to } 84.4\%$$

$$99\% \quad p \pm 2.58\sqrt{\left[\frac{p(1-p)}{n}\right]} = 0.72 \pm 2.58\sqrt{\left(\frac{0.72 \times 0.28}{50}\right)} = 55.6\% \text{ to } 88.4\%$$

As mentioned earlier, these confidence limits are approximate because of the use of the normal distribution and because of substitution of the sample proportion, p, in place of the population proportion π in the expression for the standard error of p.

Note: The standard error, and hence the size of the confidence interval, depends mainly on the *actual size* of the sample and *not*, for practical purposes, on the *proportion* which the sample is of the population. The latter usually only becomes important when it is about 20% or more. In such a case, the formula used in the example overestimates the standard error a little bit, i.e. the probability associated with the calculated interval is a little higher than stated.

Thus in this example, the 50 seeds provide the same information about the overall germination rate whether they were taken randomly from a batch of 1000 seeds or a batch of 1 000 000 seeds (or any other large number).

8. As an extension of the previous example, suppose two seedsmen, A and B, each produce large quantities of nominally the same type of seed. Under standard test conditons, out of 200 seeds from A, 180 germinate, whilst 255 germinate out of 300 from B. Has A a better germination rate than B?

Set up the null hypothesis that both germination rates are the same, i.e.

$$H_0: \pi_A = \pi_B \qquad H_1: \pi_A \neq \pi_B$$

An approximately normal test statistic can be set up (see summary table 7.1) as

$$u \simeq \frac{(p_A - p_B) - (\pi_A - \pi_B)}{\sqrt{\left[\frac{\pi_A(1-\pi_A)}{n_A} + \frac{\pi_B(1-\pi_B)}{n_B}\right]}}$$

Under the null hypothesis, $\pi_A = \pi_B$ = some value π, say, and the test statistic becomes

$$u \simeq \frac{(p_A - p_B)}{\sqrt{\left[\pi(1-\pi)\left(\frac{1}{n_A} + \frac{1}{n_B}\right)\right]}}$$

The actual value of π, however, is unknown and it is usual to replace it by its pooled sample estimate, p, obtained as a weighted average of the two sample proportions p_A and p_B, the sample sizes being the weights.

Thus

$$p = \frac{n_A p_A + n_B p_B}{n_A + n_B} = \frac{180 + 255}{500} = \frac{435}{500} = 0.87$$

$$u = \frac{0.90 - 0.85}{\sqrt{[0.87 \times 0.13(\frac{1}{200} + \frac{1}{300})]}} = \frac{0.05}{0.0307} = 1.63$$

Since this value does not exceed 1.96, numerically, there is no evidence at the 5% level of a difference between the seeds of A and B as far as germination rate is concerned.

9. Example (1) of this section was concerned with a normal variable with standard deviation of 6.0 units, this being assumed constant whatever the mean of the population. The null hypothesis was set up that the mean was 50.0 units.

(*a*) If a two-sided test of this hypothesis is carried out at the 1% level of significance, what will be the type II error, if a sample of size 16 is taken and the population mean is actually equal to

(i) 51.0 units?
(ii) 53.0 units?

(*b*) What size of sample would be necessary to reject the test hypothesis with probability 90% when the population mean is actually 48.0 units? The significance level (type I error) remains at 1%.

(*a*) (i) Figure 7.1 shows the essence of the solution. The solid distribution is how $\bar{x}$ is assumed to be distributed under the null hypothesis, the critical

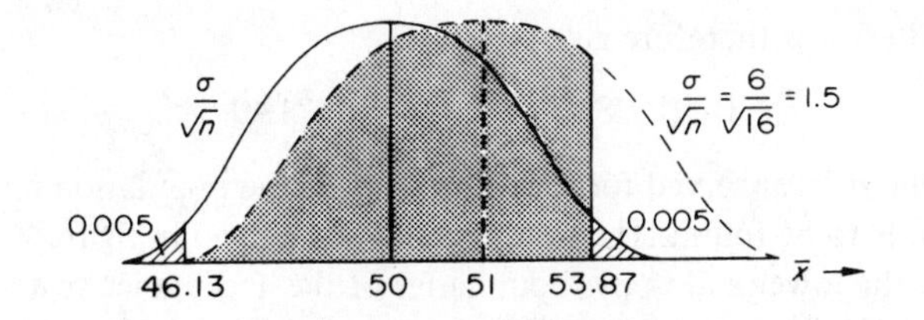

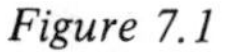
Figure 7.1

region being given by the shaded area in its two tails. The boundaries of the acceptance region for a 1% significance level are at

$$50 \pm 2.58 \times \frac{6}{\sqrt{16}} = 46.13 \text{ and } 53.87$$

The dotted distribution shows how $\bar{x}$ is actually distributed. If the observed sample mean falls in the acceptance region, the null hypothesis would not be rejected and a type II error would be committed. The shaded area shown dotted is the probability of making such an error and to find it, we need

$$u = \frac{53.87 - 51.0}{\frac{6}{\sqrt{16}}} = +1.93$$

and

$$u = \frac{46.13 - 51.0}{\frac{6}{\sqrt{16}}} = -3.25$$

The tail areas corresponding to these values are 0.0268 and 0.0006 approximately.

The type II error is therefore equal to

$$1 - (0.0268 + 0.0006) = 0.9726$$

(ii) The solution to this part is the same as that for part (i) except that the actual distribution of the sample mean will be centred around 53.0.

The values of u corresponding to the limits of the acceptance region are

$$u = \frac{53.87 - 53.0}{\frac{6}{\sqrt{16}}} = 0.58$$

and

$$u = \frac{46.13 - 53.0}{\frac{6}{\sqrt{16}}} = -4.58$$

The type II error is therefore given by

$$1-(0.2810+0.0000) = 0.7190$$

(*b*) Here the risks are fixed for specific values of the population mean; the sample size, *n*, is to be found. The requirements are shown in figure 7.2, in which $\bar{x}_1^*$ and $\bar{x}_2^*$ are the lower and upper boundaries of the acceptance region. Half per cent of the sample mean distribution assumed under the null hypothesis will lie outside each of these boundaries (1% type I error with a two-sided alternative).

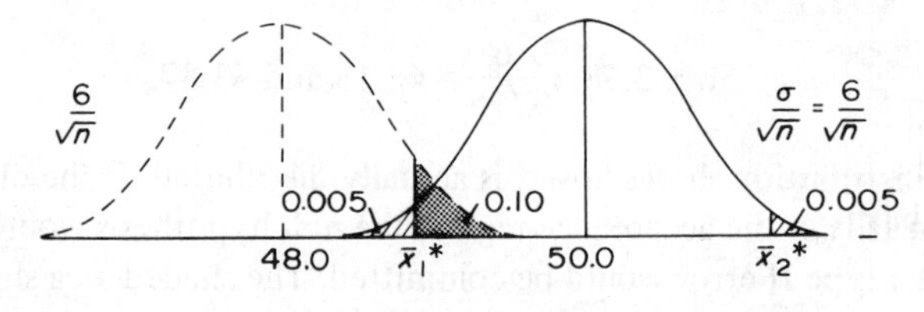

Figure 7.2

The dotted distribution shows how the means of samples of size *n* will be distributed when the population average is actually 48.0 units. The extreme part of the right-hand tail of this distribution will lie above $\bar{x}_2^*$ but it will be such a minute proportion in this case as to be negligible.

The following equations can be set up.

$$\bar{x}_1^* = 48.0 + 1.28 \times \frac{6}{\sqrt{n}} \tag{7.1}$$

$$\bar{x}_1^* = 50.0 - 2.58 \times \frac{6}{\sqrt{n}} \tag{7.2}$$

Subtracting one equation from the other leads to

$$(1.28 + 2.58)\frac{6}{\sqrt{n}} = (50.0 - 48.0)$$

or

$$n = \left(\frac{3.86 \times 6}{2}\right)^2 = 11.58^2 = 134$$

The critical values of $\bar{x}$ are thus

$$50.0 \pm 2.58 \times \frac{6}{\sqrt{134}} = 48.66 \text{ and } 51.34$$

Part (*b*) of this example postulated the requirement that if the mean is 48.0 units (or more generally, *if* it differs from the test value by more than 2.0 units), the chance of detecting such a difference should be 90%. This requirement would have been determined by the practical aspects of the problem. However, if the actual population mean were less than 48.0 (or bigger than 52.0), the probability of committing a type II error with a sample size of 134 would be less than 10%; and if the population mean were actually between 48.0 and 50.0, this probability would be greater than 10%.

10. What is the smallest random sample of seeds necessary for it to be asserted, with a probability of at least 0.95, that the observed sample germination proportion deviates from the population germination rate by less than 0.03?

The standard error of a sample proportion is $\sqrt{[\pi(1-\pi)/n]}$ where π is the population proportion and n the sample size. Assuming that n will be large enough for the approximation of normality to apply reasonably well to the distribution of p, the problem requires that

$$1.96\sqrt{\left[\frac{\pi(1-\pi)}{n}\right]} = 0.03$$

giving

$$n = \left(\frac{1.96}{0.03}\right)^2 \pi(1-\pi)$$

π, the quantity to be estimated is unknown (if it were known, there would be no need to estimate it) and this creates a slight difficulty in determining n. However, $\pi(1-\pi)$ takes its maximum value of $\frac{1}{4}$ when $\pi = \frac{1}{2}$ and

$$n = \left(\frac{1.96}{0.03}\right)^2 \tfrac{1}{4} \simeq 1060$$

would certainly satisfy the conditions of the problem (whatever the value of π).

Alternatively if an estimate is available of the likely value of π, this can be used instead of π as an approximation. Such an estimate may come from previous experience of the population or perhaps from a pilot random sample; the pilot sample estimate can be used to determine the total size necessary. If the pilot sample is at least as big as this, no further sampling is needed. If it was not, the *extra* number of observations required can be found approximately. If such extra sampling is not possible for some reason (too costly, not enough time), the confidence probabilities of types I and II errors will be modified (adversely).

For this example, if the seed population germination rate is usually about 80%, then the required value of sample size for at most a deviation of 0.03 (i.e. 3%) with probability of 0.95 is

$$n = \left(\frac{1.96}{0.03}\right)^2 0.8 \times 0.2 \simeq 680$$

(c.f. 1060 before).

7.3 Problems for Solution

1. In production of a tinned product, records show that the standard deviation of filled weights is 0.025 kg. A sample of six tins gave the following weights: 1.04, 0.97, 0.99, 1.00, 1.02, 1.01 kg.

(*a*) If the process is required to give an average weight of 1.00 kg does the filling machine require re-setting?

(*b*) Determine confidence limits for the actual process average.

2. In a dice game, if an odd number appears you pay your opponent 1p and if an even number turns up, you receive 1p from him. If, after 200 throws, you are losing 50p and the dice are your opponent's, would you be justified in feeling cheated?

3. A company, to determine the utilisation of one of its machines, makes random spot checks to find out for what proportion of time the machine is in use. It is found to be in use during (*a*) 49 out of 100 checks, and (*b*) 280 out of 600 checks.

Find out in each case, the percentage time the machine is in use, stating the confidence limits.

How many random spot checks would have to be made to be able to estimate the machine utilisation to within ± 2%?

4. In a straight election contest between two candidates, a survey poll of 2000 gave 1100 supporting candidate A. Assuming sample opinion to represent performance at the election, will candidate A be elected?

5. In connection with its marketing policy, a firm plans a market research survey in a country area and another survey in a town. A random sample of the people living in the areas is interviewed and one question they are asked is whether or not they use a product of the firm concerned. The results of this question are:

Town:	Sample size = 2000, no. of users = 180
Country:	Sample size = 2000, no. of users = 200

Does this result show that the firm's product is used more in the country than in town?

6. In a factory, sub-assemblies are supplied by two sub-contractors. Over a period, a random sample of 200 from supplier A was 5% defective, while a sample of 300 from supplier B was 3% defective.

Does this signify that supplier B is better than supplier A?

A further sample of 400 items from B contained eight defective sub-assemblies. What is the position now?

7. If men's heights are normally distributed with mean of 1.73 m and standard

deviation of 0.076 m and women's heights are normally distributed with mean of 1.65 m and standard deviation of 0.064 m, and if, in a random sample of 100 married couples, 0.05 m was the average value of the difference between husband's height and wife's height, is the choice of partner in marriage influenced by consideration of height?

8. For the data of problem (3) (page 46), chapter 2, estimate 99% confidence limits for the mean time interval between customer arrivals. Also find the number of observations necessary to estimate the mean time to within 0.2 min.

9. An investigation of the relative merits of two kinds of electric battery showed that a random sample of 100 batteries of brand A had an average lifetime of 24.2 h, with a standard deviation of 1.8 h, while a random sample of 80 batteries of brand B had an average lifetime of 24.5 h with a standard deviation of 1.5 h.

Use a significance level of 0.01 to test whether the observed difference between the two average lifetimes is significant.

10. Two chemists, A and B, each perform independent repeat analyses on a homogeneous mixture to estimate the percentage of a given constituent.

The repeatability of measurement has a standard deviation of 0.1% and is the same for each analyst. Four determinations by A have a mean of 28.4% and five readings by B have a mean of 28.2%.

(*a*) Is there a systematic difference between the analysts?

(*b*) If each analyst carries out the same number of observations as the other, what should this number be in order to detect a systematic difference between the analysts of 0.3% with probability of at least 99%, the level of significance being 1%?

7.4 Solutions to Problems

1. The observed sample mean is $\bar{x} = (1.04 + 0.97 + 0.99 + 1.00 + 1.02 + 1.01)/6$

$$= 1.005 \text{ kg}$$

(*a*) Assuming the mean net weight of individual cans is 1.00 kg, i.e.

$$H_0: E[x] = \mu_0 = 1.00 \text{ kg}$$

$$H_1: E[x] = \mu_1 \neq 1.00 \text{ kg}$$

then

$$u = \frac{\bar{x} - \mu_0}{\sigma/\sqrt{n}} = \frac{1.005 - 1.000}{0.025/\sqrt{6}} = \frac{\sqrt{6}}{5} = 0.49$$

The probability of such a deviation is about 62% and so there is no real evidence that the process average is not 1.00 kg, i.e. the sample data are quite consistent with a setting of 1.00 kg, although a type II error could be committed in deciding not to re-set the process.

(*b*) Confidence limits for the actual current process average are, for two levels of confidence

$$95\%: \quad 1.005 \pm 1.96 \times \frac{0.025}{\sqrt{6}} = 1.005 \pm 0.02 = 0.985 \text{ and } 1.025 \text{ kg}$$

$$99\%: \quad 1.005 \pm 2.58 \times \frac{0.025}{\sqrt{6}} = 1.005 \pm 0.026 = 0.979 \text{ and } 1.031 \text{ kg}$$

2. Losing 50p in 200 throws means that there must have been 125 odd numbers (losing results) and 75 even numbers (winners) in 200 throws. Set up the null hypothesis that the dice is unbiased.

$$H_0: \pi = 0.5 \ (\pi \equiv \text{the probability of an odd number})$$

$$H_1: \pi \neq 0.5$$

The total number of odd numbers will be binomially distributed and since $n = 200$ and $\pi = \frac{1}{2}$ we know that

$$u \simeq \frac{x - n\pi}{\sqrt{[n\pi(1-\pi)]}} = \frac{124.5 - 100}{\sqrt{(200 \times \frac{1}{2} \times \frac{1}{2})}} \text{ making the correction for continuity}$$

$$= \frac{24.5}{\sqrt{50}} = 3.46$$

The probability of such a deviation is certainly less than 0.0007 and it therefore seems likely that the dice is biased towards odd numbers.

3. 95% confidence limits for the proportional utilisation of the machine are approximately

$$p \pm 1.96\sqrt{\left[\frac{p(1-p)}{n}\right]}$$

which gives

$$(a) \ 0.49 \pm 1.96\sqrt{\left(\frac{0.49 \times 0.51}{100}\right)} = 0.49 \pm 0.098 = 39.2 \text{ to } 58.8\%$$

and

$$(b) \ \frac{280}{600} \pm 1.96\sqrt{\left(\frac{280 \times 320}{600 \times 600 \times 600}\right)} = 0.467 \pm 0.04 = 42.7 \text{ to } 50.7\%$$

Note: The standard error has been reduced by a factor of approximately $\sqrt{6}$, the square root of the ratio of the two sample sizes.

Also, since π is near to 0.5, for 95% confidence estimation, the required number of spot checks is given by

$$1.96\sqrt{\left(\frac{0.5 \times 0.5}{n}\right)} = 0.02, \quad \text{i.e.} \quad n = 98^2 \times \tfrac{1}{4} = 2401$$

For a 99% confidence interval of width (2 x 0.02), the required n is found from

$$2.58\sqrt{\left(\frac{0.5 \times 0.5}{n}\right)} = 0.02 \qquad n = 129^2 \times \tfrac{1}{4} = 4160$$

4. 99% confidence limits for the population proportional support for candidate A are

$$\frac{1100}{2000} \pm 2.58\sqrt{\left(\frac{1100 \times 900}{2000 \times 2000 \times 2000}\right)} = 0.55 \pm 0.0287 = 0.521 \text{ to } 0.579$$

Thus it is virtually certain that candidate A will be elected.

5. Assume that there is no difference in the proportion of people using the product either in the country (π_C) or in the town (π_T).

$$H_0: \pi_C = \pi_T \qquad H_1: \pi_C \neq \pi_T$$

The best estimate under H_0 of the common usage rate

$$= \frac{200 + 180}{4000} = 0.095$$

Then

$$u \simeq \frac{(0.10 - 0.09) - 0}{\sqrt{[0.095 \times 0.905\,(\frac{1}{2000} + \frac{1}{2000})]}} = 1.08$$

There is no evidence that the proportion of people in the country area using the product is any different from that in the town.

6. Assume that the percentage of sub-assemblies which are defective is the same in the long run for both suppliers.

Thus

$$H_0: \pi_A - \pi_B = 0 \qquad H_1: \pi_A - \pi_B \neq 0$$

Assuming H_0 to be true, the best estimate of each supplier's defective proportion is

$$\hat{\pi} = \frac{200 \times 0.05 + 300 \times 0.03}{200 + 300} = \frac{19}{500}$$

Thus

$$u \simeq \frac{(0.05-0.03)-0}{\sqrt{[\frac{19}{500}\times\frac{481}{500}(\frac{1}{200}+\frac{1}{300})]}} = \frac{0.02\times 500}{\sqrt{(19\times 481\times\frac{5}{600})}} = 1.145$$

There is no evidence of a difference between the suppliers.

With the additional evidence, assuming that the underlying conditions remain unchanged, the test may be carried out again

$$H_0: \pi_A - \pi_B = 0 \qquad H_1: \pi_A - \pi_B \neq 0$$

Pooling all the information,

$$\hat{\pi} = \frac{10+9+8}{200+300+400} = \frac{27}{900} = 0.03$$

$$u \simeq \frac{(\frac{10}{200}-\frac{17}{700})-0}{\sqrt{[0.03\times 0.97(\frac{1}{200}+\frac{1}{700})]}} = 1.88$$

This value of u nearly reaches its critical value for a 5% (two-sided) level of significance; the actual level is about 6%. There is thus some suspicion that B is better than A but what action is taken depends on the consequences of the possible alternative decisions.

7. Set up the test hypothesis that the choice of marriage partner is not influenced by the height of either. In this case, in a married couple, the height of a man and of a woman is each a random selection from the distributions of men's and women's heights respectively.

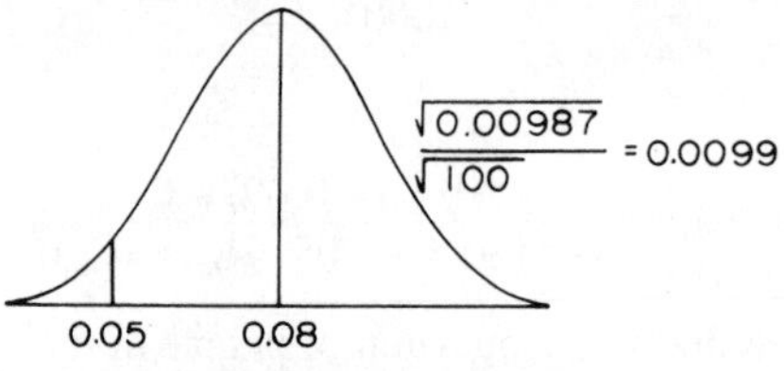

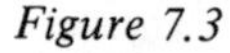

Figure 7.3 Average height excess for 100 couples

The excess of the man's height over the woman's height will be a normal variable with mean of (1.73 – 1.65) m and variance of $(0.076^2 + 0.064^2)\text{m}^2$. The average difference (excess) of height taken over a random sample of 100 such married couples will be normally distributed (i.e. from one sample of 100 to another) with mean of 1.73 – 1.65 = 0.08 m and variance of $(0.076^2 + 0.064^2)/100$ and have a standard error of $\sqrt{0.00987}/\sqrt{100} = 0.0099$ m.

The observed average difference was 0.05 m corresponding to a u value of

$$\frac{0.05-0.08}{0.0099} = -3.04$$

The (two-sided) significance level corresponding to this is approximately 0.0024 and thus it seems reasonable to conclude that the choice of marriage partner is not independent of height.

8. The observed data of problem 3, chapter 2, are distributed in a skew pattern with a calculated mean of 1.29 min and standard deviation of 1.14 min. The figure of 1.29 min is the average of 56 individual readings, and by the central limit theorem, such an average can be expected to be normally distributed. The appropriate confidence limits can be found using the sample standard deviation as an estimate of the population standard deviation since it is based on more than 30 readings; such an approximation will be good enough for most practical purposes.

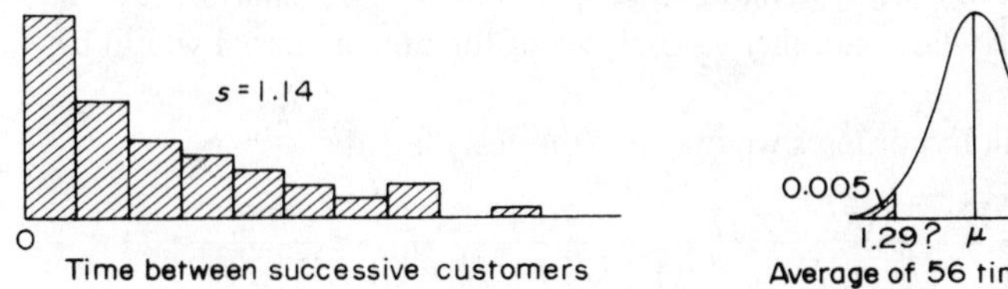

Figure 7.4

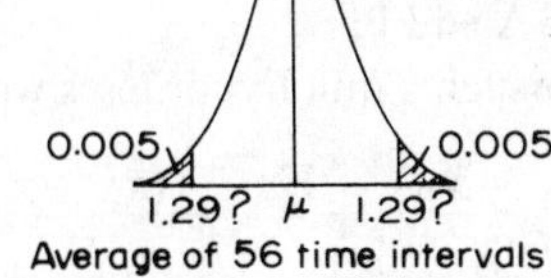

Figure 7.5

99% confidence limits for the mean time between arrivals are

$$1.29 \pm 2.58 \times \frac{1.14}{\sqrt{56}} = 1.29 \pm 0.39 = 0.90 \text{ and } 1.68 \text{ min}$$

The number of observations, n, necessary to estimate the population mean to within 0.2 min (99% confidence) is given by equating the sampling error to the required error, i.e.

$$2.58 \frac{\sigma}{\sqrt{n}} = 0.2 \qquad n = \left(\frac{2.58 \times 1.14}{0.2}\right)^2 = 216$$

9. Set up the test hypothesis that there is no difference in mean lifetimes between the two brands.

$$H_0: E[(\bar{x}_A - \bar{x}_B)] = \mu_A - \mu_B = 0$$

$$H_1: E[(\bar{x}_A - \bar{x}_B)] = \mu_A - \mu_B \neq 0$$

An appropriate statistic is

$$u = \frac{(\bar{x}_A - \bar{x}_B) - 0}{\sqrt{\left(\frac{\sigma_A^2}{n_A} + \frac{\sigma_B^2}{n_B}\right)}}$$

The denominator being the standard error of the difference of two sample means based on samples of size n_A and n_B respectively.

Thus

$$u = \frac{(24.2 - 24.5) - 0}{\sqrt{\left(\frac{1.8^2}{100} + \frac{1.5^2}{80}\right)}}$$

substituting the sample variances for the population variances

$$= \frac{-0.3}{\sqrt{0.0605}} = \frac{-0.3}{0.246} = -1.22$$

Since this value is not numerically larger than 2.58, there is no evidence of a difference in mean lifetimes between A and B.

10. (*a*) Assume there is no systematic difference between the analysts, i.e. the means of an infinitely large number of analyses of the same material would be equal for A and B.

Under such a null hypotheses we may use the test statistic

$$u = \frac{(\bar{x}_A - \bar{x}_B) - 0}{\sqrt{\left(\frac{\sigma_A^2}{n_A} + \frac{\sigma_B^2}{n_B}\right)}} = \frac{(\bar{x}_A - \bar{x}_B)}{\sqrt{\left[\sigma^2\left(\frac{1}{n_A} + \frac{1}{n_B}\right)\right]}} = \frac{28.4 - 28.2}{\sqrt{[0.1^2(\frac{1}{4} + \frac{1}{5})]}} = \frac{0.2}{0.1\sqrt{\frac{9}{20}}} = 2.98$$

This is significant at the 1% level (i.e. $|u| > 2.58$) and we can conclude (with only a small type I error) that there is a systematic difference between the analysts, A giving a higher result than B on average. Thus at least one of them, and possibly both, gives a biased estimate of the actual percentage composition.

99% confidence limits for the extent of this systematic difference are given by

$$(28.4 - 28.2) \pm 2.58\sqrt{[0.1^2(\tfrac{1}{4} + \tfrac{1}{5})]} = 0.2 \pm 0.173 = 0.027 \text{ and } 0.373\%$$

(*b*) Figure 7.o shows the requirements of this problem.

$$(\bar{x}_A - \bar{x}_B)^* = 0 + 2.58 \times \sqrt{\left[\sigma^2\left(\frac{1}{n_A} + \frac{1}{n_B}\right)\right]} \tag{7.3}$$

$$(\bar{x}_A - \bar{x}_B)^* = 0.3 - 2.33\sqrt{\left[\sigma^2\left(\frac{1}{n_A} + \frac{1}{n_B}\right)\right]} \tag{7.4}$$

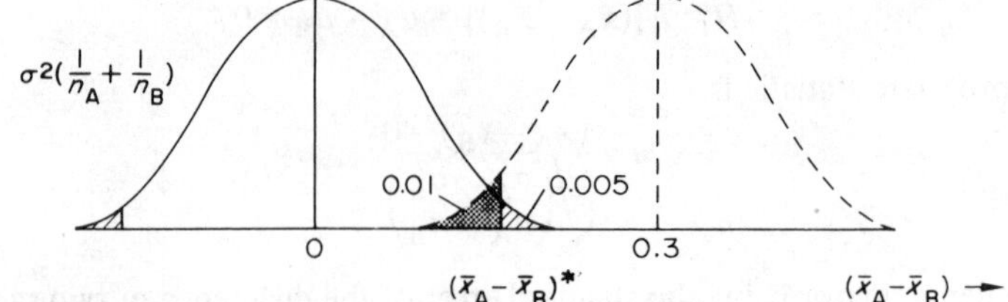

Figure 7.6

An equivalent pair of equations would be obtained for a systematic difference of –0.3%.

Note: In writing down equations (7.3) and (7.4), the minute part of the left-hand tail of the dotted distribution falling in the lower part of the critical region has been ignored.

Putting $n_A = n_B = n$ gives the required number of readings by each analyst as

$$n = \frac{(2.58 + 2.33)^2 \times 2 \times 0.1^2}{0.3^2} = 4.91^2 \times \tfrac{2}{9} = 5.35$$

Thus each analyst should do six tests, the probability of detecting a systematic difference of 0.3% between them (if it exists) being greater than the required minimum of 99%.

In fact the required minimum power would still be achieved if one analyst took six tests and the other five in order to reduce the total cost or effort involved.

7.5 Practical Laboratory Experiments and Demonstrations

Since this chapter is concerned with 'large sample' methods, all the experiments and demonstrations on illustrating the basic concepts of significance have been left over to chapter 8.

In view of sample sizes required, experimenting is not very effective for methods outlined in this chapter.

8 Sampling theory and significance testing (II)—'t', 'F' and χ^2 tests

8.1 Syllabus Covered

Unbiased estimate of population variance; degrees of freedom; small sampling theory; '*t*' test of significance; confidence limits using '*t*'; paired comparisons; '*F*' test of significance for two variances; χ^2 test of significance; goodness of fit tests; contingency tables.

8.2 Resume of Theory and Basic Concepts

8.2.1 *Unbiased Estimate of Population Variance*

In chapter 7 the use of significance testing for large samples or for samples where an independent estimate of population variance was available was discussed. The '*u*' test was described for comparing a sample mean with a given hypothesis and also for testing significant differences between two population means.

In this chapter the problems outlined are different in that the sample sizes are small and no independent estimate of population variance is available—an estimate from the sample having to be used for the population variance.

In obtaining an unbiased estimate of the population variance from sample data the following formula must be used

$$s^2 = \frac{\sum_{i}^{n} (x_i - \bar{x})^2}{n-1} \tag{8.1}$$

where $\bar{x}$ = sample average.

Note: If an independent estimate of population mean μ is available the sample estimator of variance is

$$s^2 = \frac{\sum_{i}^{n} (x_i - \mu)^2}{n} \tag{8.2}$$

The denominators in both equations (8.1) and (8.2) are called the degrees of freedom of the variance estimate.

8.2.2 *Degrees of Freedom*

This concept of degrees of freedom is very difficult to define exactly but it can be considered as the number of independent variates. This number of independent variates or degrees of freedom is equal to the total number of variates less the number of independent linear constraints on the variates.

For example in equation (8.1) in estimating the population variance, the sample mean $\bar{x}$ is used in the equation thus reducing the number of comparisons or degrees of freedom to $n - 1$. No such reduction is necessary in equation (8.2). When dealing with χ^2 goodness of fit testing a further explanation of this concept of degrees of freedom will be given.

8.2.3 *The 'u' Test with Small Samples*

The arbitrary division of significance testing between large sampling theory (or approximate methods) in chapter 7 and the small sampling theory (or exact methods) in this chapter, necessitates the repeating of one test in order to maintain consistency.

Testing the Hypothesis that the Mean of a Normal Population has a Specific Value μ_0–Population Variance Known

Here, providing the population variance is known (and therefore the sample estimate of variance is not used), then the '*u*' test is appropriate whatever the sample size.

Thus

$$u = \frac{\bar{x} - \mu_0}{\dfrac{\sigma}{\sqrt{n}}}$$

is calculated and the significance level is determined.

Example

In an intelligence test on ten pupils the following scores were obtained: 105, 120, 90, 85, 130, 110, 120, 115, 125, 100.

Given that the average score for the class before the special tuition for the test was 105 with standard deviation 8.0, has the special tuition improved the performance?

Here since the standard deviation is given and if the assumption is made that tuition method does not change this variation, then the *u* test is applicable.

Null hypothesis–tuition has made no improvement

Average score in test

$$\bar{x} = \frac{105 + 120 + \ldots + 125 + 100}{10} = 110$$

Here a one-tailed test can be used if it is again assumed that tuition could not have worsened the performance.

Thus

$$u = \frac{110 - 105}{\frac{8}{\sqrt{10}}} = 1.98$$

From table 3*

$$\text{for } 5\% \quad u = 1.64$$

$$1\% \quad u = 2.33$$

This result is significant at the 5% level; there is evidence of an improvement.

8.2.4 *The 't' Test of Significance*

Testing the Hypothesis that the Mean of a Normal Population has a Specific Value μ_0–Population Variance Unknown

Here the sample of size n is used to give the estimate of population variance.

$$s^2 = \frac{\Sigma(x_i - \bar{x})}{n-1} = \frac{\Sigma x_i^2 - \frac{(\Sigma x_i)^2}{n}}{n-1} \tag{8.3}$$

The null hypothesis is set up that the sample has come from a normal population with mean μ_0.

W. S. Gosset under the nom de plume of 'Student' examined the following statistic

$$t = \frac{\bar{x} - \mu_0}{\frac{s}{\sqrt{n}}} \tag{8.4}$$

and showed that it is not distributed normally but in a form which depends on the degrees of freedom (v) if the null hypothesis is true. Table 7* sets out the various percentage points of the 't' distribution for a range of degrees of freedom. Obviously t tends to the statistic u in the limit where $v \to \infty$, i.e. t is approximately normally distributed for large degrees of freedom v. Reference to the table* shows that, as most textbooks assert, where the degrees of freedom exceed 30, the normal approximation can be used or the 't' test can be replaced by the simpler 'u' test.

Note: For a two-tailed test note that a 5% significance level requires $\alpha = 0.025$ in table 7* and for a 1% significance level, $\alpha = 0.005$ (α is a proportion, not a percentage). See section 8.2.6.

Again, one-tailed tests are only used when *a priori* logic clearly shows that the alternative population mean must be on one side of the hypothesis value μ_0. See section 8.2.7.

Testing the Hypothesis that the Means of Two Normal Populations are μ_x and μ_y Respectively–Variances Equal but Unknown

Note: The assumption must hold that the variances of the two populations are the same (i.e. $\sigma_x^2 = \sigma_y^2$) since we are going to pool two sample variances and this only makes sense if they are both estimates of the same thing–a common population variance. If σ_x^2 does not equal σ_y^2 then the statistic given below is not distributed like t.

The two sample variances s_x^2 and s_y^2 are pooled to give a best estimate of the common population variance.

$$s^2 = \frac{(n_x - 1)\, s_x^2 + (n_y - 1)\, s_y^2}{n_x + n_y - 2} \tag{8.5}$$

where n_x and n_y are the sizes of the two samples, and

$$t = \frac{(\bar{x} - \bar{y}) - (\mu_x - \mu_y)}{s\sqrt{\left(\frac{1}{n_x} + \frac{1}{n_y}\right)}} \tag{8.6}$$

with $(n_x + n_y - 2)$ degrees of freedom. The usual test hypothesis is that the populations have equal means and under this assumption $(\mu_x - \mu_y) = 0$ and the test statistic reduces to

$$t = \frac{(\bar{x} - \bar{y})}{s\sqrt{\left(\frac{1}{n_x} + \frac{1}{n_y}\right)}} \tag{8.7}$$

t-Test Using Paired Comparisons

In many problems, the power of the significance test can be increased by pairing the results and testing the hypothesis that the mean difference between paired readings is equal to μ_0.

Note: This approach is only legitimate provided that there is a valid reason for pairing the observations. This validity is determined by the way in which the experimental observations are obtained.

Let the number of paired readings $= n$

Let the difference of the ith pair $= d_i$

Then

$$s^2 = \frac{\sum_i^n (d_i - \bar{d})^2}{n-1} \quad \text{where} \quad \bar{d} = \frac{\sum_i^n d_i}{n}$$

and

$$t = \frac{\bar{d} - \mu_0}{\frac{s}{\sqrt{n}}} \tag{8.8}$$

The test hypothesis is usually of the null type where there is assumed to be no difference on average in the paired readings, i.e. $\mu_0 = 0$. In this case the test statistic t is given by

$$t = \frac{\bar{d}}{\frac{\hat{s}}{\sqrt{n}}} \tag{8.9}$$

Confidence Limits for Population Mean

Where the degrees of freedom are less than about 30, the confidence limits for population mean μ_0 are:

for 95% confidence limits $\bar{x} \pm (t_{0.025,\upsilon}) \dfrac{s}{\sqrt{n}}$

for 99% confidence limits $\bar{x} \pm (t_{0.005,\upsilon}) \dfrac{s}{\sqrt{n}}$

This is similar to the large sample case except that t is used instead of u.

8.2.5 *The 'F' Test of Significance*

For testing the hypothesis that the variances of two normal populations are equal.

Again, a null hypothesis is set up that the variances are the same.

Let

$$s_x^2 = \frac{\Sigma(x_i - \bar{x})^2}{n_x - 1} \qquad s_y^2 = \frac{\Sigma(y_i - \bar{y})^2}{n_y - 1}$$

Then

$$F = \frac{\hat{s}_x^2}{\hat{s}_y^2} \quad \text{where} \quad \hat{s}_x^2 > \hat{s}_y^2 \tag{8.10}$$

If F is greater than $F_{0.025}$ (see table 9*) for $(n_x - 1)$ degrees of freedom of numerator and $(n_y - 1)$ degrees of freedom for the denominator, then the difference is significant at 5% level ($\alpha = 0.05$). For F to be significant at 1% level, use $F_{0.005}$ (actually $F_{0.01}$ will have to be used giving a 2% significance level of F).

8.2.6 *The χ^2 Test of Significance*

Definition of χ^2

Let $x_1, x_2, \ldots x_n$ be n normal variates from a population with mean μ and standard deviation σ.

Then

$$\chi_n^2 = \left(\frac{x_1 - \mu}{\sigma}\right)^2 + \left(\frac{x_2 - \mu}{\sigma}\right)^2 + \ldots + \left(\frac{x_n - \mu}{\sigma}\right)^2 \qquad (8.11)$$

the suffix n denoting the number of degrees of freedom. Obviously the larger n is the larger χ^2 and the percentage points of the sampling distribution of χ^2 are given in table 8*.

For example, where $n = 1$ the numerical value of the standardised normal deviate u exceeds 1.96 with 5% probability and 2.58 with 1% probability (i.e. with half the probabilities in each tail). Consequently χ^2 with one degree of freedom has 5% and 1% points as 1.96^2 and 2.58^2 or 3.841 and 6.635.

However, for higher degrees of freedom the distribution of χ^2 is much more difficult to calculate, but it is fully tabulated in table 8*.

Goodness of Fit Test using χ^2

A most important use of the χ^2 distribution is in a significance test for the 'goodness of fit' between observed data and an hypothesis.

Let k = number of cells or comparisons

O_i = observed frequency in ith cell

E_i = expected frequency in ith cell from the hypothesis

r = number of restrictions, *derived from the observed readings,* which have to be used when fitting the hypothesis.

Then

$$\sum_{i=1}^{n} \frac{(O_i - E_i)^2}{E_i}$$

is distributed like χ^2 with $(k - r)$ degrees of freedom where r = number of parameters used to fit the distribution.

For the use of this test all the E_i values must be greater than 5. If any are less then the data must be grouped.

Application of χ^2 to Contingency Tables

When the total frequency can be divided between two factors, and each factor subdivided into various levels, then the table formed is called a contingency table. Data in the form of a contingency table give one of the simplest methods of testing the relationship between two factors.

Consider the following contingency table (table 8.1) with the first factor (F_1) at a levels and the second factor (F_2) at b levels. The individual cell totals O_{ij} give the observed frequency of readings at the ith level of factor F_1 and the jth level of factor F_2.

	Factor 1 1	2	3	...	i	...	a	Row totals
Factor 2								
1	O_{11}	O_{21}	O_{31}		O_{i1}		O_{a1}	$\sum_i O_{i1}$
2	O_{12}	O_{22}						
3	O_{13}							
$\vdots$								
j	O_{1j}				O_{ij}			$\sum_i O_{ij}$
$\vdots$								
b	O_{1b}						O_{ab}	$\sum_i O_{ib}$
Column totals	$\sum_j O_{1j}$				$\sum_j O_{ij}$		$\sum_j O_{aj}$	$\sum_i \sum_j O_{ij}$

Table 8.1

$\sum_i \sum_j O_{ij}$ = total frequency

$\sum_j O_{ij}$ = total frequency at the ith level of factor 1 (column total)

$\sum_i O_{ij}$ = total frequency at the jth level of factor 2 (row total)

These tables are generally used to test the hypothesis that the factors are independent.

If this hypothesis is true then, the expected cell frequency

$$E_{ij} = \frac{\sum_i O_{ij} \times \sum_j O_{ij}}{\sum_i \sum_j O_{ij}} \quad \text{and} \quad \sum_i \sum_j \left[\frac{(O_{ij} - E_{ij})^2}{E_{ij}}\right]$$

is distributed as χ^2 with $(a-1)(b-1)$ degrees of freedom.

It can be shown that only $(a-1)(b-1)$ of the comparisons are independent since the row and column totals of expected frequencies must be the same as the row and column totals of observed frequencies.

8.2.7 *One- and Two-tailed Tests*

This whole question of one- and two-tailed tests is a subject of considerable controversy among statisticians.

However, the following points of guidance are useful in deciding which to apply.

(1) In general, if ever in doubt use the two-tailed test since this plays safer.

(2) Only if, from *a priori* knowledge, it can be definitely stated that the change must be in one direction only, can the one-tailed test be used.

The observations apply, of course, to all significance tests and it is hoped that the examples given will clarify this confusing problem.

8.2.8 *Examples on the Use of the Tests*

1. A canning machine is required to turn out cans weighing 251 g on the average. A random sample of five is drawn from the output and each is found to weigh 252 g respectively. Can it be said that the machine produces cans of average weight 251 g?

Coding the variate by subtracting 250 gives $x = 1, 2, 4, 4, 2$: $\Sigma x = 13$, $\Sigma x^2 = 41$.

Null hypothesis is set up that the process is running at an average of 251 g.

Mean

$$\bar{x} = \frac{13}{5} = 2.6$$

Estimated variance of population

$$s = \frac{\Sigma x^2 - \frac{(\Sigma x)^2}{n}}{n-1} = 1.8$$

$\therefore$ Estimated standard deviation of population $s = 1.34$

$\therefore$ Estimated standard error of sample mean $\epsilon_{\bar{x}} = \dfrac{1.34}{\sqrt{5}} = 0.6$

On the null hypothesis that the population average is 1

$$t = \frac{2.6 - 1}{0.6} = 2.67$$

From tables* (4 degrees of freedom)

$$t_{0.025} = 2.78 \qquad t_{0.005} = 4.6$$

or the results are not significant. However, since the t value is close to the 5% level (2-tailed) it is possible that if a large sample were taken a difference may be shown.

2. A weaving firm has been employing two methods of training weavers. The first is the method of 'accelerated training', the second, 'the traditional' method. Although it is accepted that the former method enables weavers to be trained more quickly, it is desired to test the long-term effects on weaver efficiency. For this purpose the varying efficiency of the weavers who have undergone training during a period of years has been calculated, and is given in table 8.2.

Is there any significant difference between training methods?

		Specialised training A	Traditional method B	Total
Above shed average	1	32	12	44
Below shed average	2	14	22	36
Insufficient data	3	6	9	15
Total		52	43	95

Table 8.2. Training schemes and weaver efficiency

Null hypothesis set up that there is no difference in the methods.

$$\therefore \quad E_{A1} = \tfrac{52}{95} \times 44 = 24.1 \qquad E_{B1} = 44 - 24.1 = 19.9$$

$$E_{A2} = \tfrac{52}{95} \times 36 = 19.7 \qquad E_{B2} = 36 - 19.7 = 16.3$$

$$E_{A3} = \tfrac{52}{95} \times 15 = 8.2 \qquad E_{B3} = 15 - 8.2 = 6.8$$

$$\chi^2 = \frac{7.9^2}{24.1} + \frac{7.9^2}{19.9} + \frac{5.7^2}{19.7} + \frac{5.7^2}{16.3} + \frac{2.2^2}{8.2} + \frac{2.2^2}{6.8}$$

$$= 2.59 + 3.14 + 1.65 + 2.00 + 0.59 + 0.79$$

$$= 10.76$$

Degrees of freedom = $(3-1)(2-1) = 2$

$$\chi^2_{0.05} = 5.991 \qquad \chi^2_{0.01} = 9.210$$

		Special Training A	Traditional B	Total
Above average	1	O = 32 E = 24.1	O = 12 E = 19.9	44
Below average	2	O = 14 E = 19.7	O = 22 E = 16.3	36
Insufficient data	3	O = 6 E = 8.2	O = 9 E = 6.8	15
Total		52	43	95

Table 8.3

Result is significant at 1% level.

∴ There is evidence that the training methods differ in their long-term efficiency.

3. In a study of two processes in industry the following data were obtained.

	Process 1	Process 2
Sample size	50	60
Mean	10.2	11.1
Standard deviation	2.7	2.1

Table 8.4

Is there any evidence of a difference in variability between the processes?

A further sample taken on process 1, gave: sample size = 100, mean = 10.6, standard deviation = 3.1.

What is the significance of the difference in variability now?

Null hypothesis set up is that there is no difference in the variation of the two processes.

$$F = \frac{\text{Greater estimate of population variance}}{\text{Lesser estimate of population variance}} = \frac{2.7^2}{2.1^2} = 1.65$$

Referring to table 9*

Degrees of freedom of greater estimate $\upsilon_1 = 49$, read as 24 (safer than ∞)

Degrees of freedom of lesser estimate $\upsilon_2 = 59$, read as 60

To be significant F must reach 1.88 at 5% level or 2.69 at 0.2% level.

∴ The difference is not significant.
Variance of combined sample for process 1

$$= \frac{s_1^2(n_1 - 1) + s_2^2(n_2 - 1)}{n_1 + n_2 - 2} = \frac{2.7^2 \times 49 + 3.1^2 \times 99}{50 + 100 - 2}$$

$$= \frac{357 + 951}{148} = 8.83$$

$$F = \frac{8.83}{2.1^2} = 2.00$$

From table 9*

Degrees of freedom of greater estimate = 149, read as ∞

Degrees of freedom of lesser estimate = 59, read as 60

5% significance level = 1.48

0.2% significance level = 1.89

Difference is highly significant or there is strong evidence that process variations are different.

4. For example (1), page 69, in chapter 3, for goals scored per soccer match, test whether this distribution agrees with the Poisson law.
Null hypothesis: the distribution agrees with the Poisson law.

No. of goals/match	0	1	2	3	4	5	6	7	8	Total
Actual frequency (O)	2	9	11	15	8	5	5	1	1	57
(grouped)	11						7			
Poisson frequency (E)	2.6	8.0	12.4	12.7	9.9	6.1	3.2	1.4	0.8	57
(grouped)	10.6						5.4			

Table 8.5

In table 8.5 the last three class intervals must be grouped to give each class interval an expected value greater than 5. Also, the first two.

$$\therefore \quad \chi^2 = \frac{(11-10.6)^2}{10.6} + \frac{(11-12.4)^2}{12.4} + \ldots + \frac{(5-6.1)^2}{6.1} + \frac{(7-5.4)^2}{7} = 1.63$$

Degrees of freedom = 6–1–1 = 4, since the totals are made the same and the Poisson distribution is fitted with same mean as the actual distribution.

Referring to table 8*

$$\chi^2_{0.05} = 9.488 \qquad \chi^2_{0.01} = 13.277$$

Thus, there is no evidence from the data for rejecting the hypothesis, or the pattern of variation shows no evidence of not having arisen randomly.

5. In a mixed sixth form the marks of eight boys and eight girls in a subject were

Boys: 25, 30, 42, 44, 59, 73, 82, 85; boys' average = 55

Girls: 32, 36, 40, 41, 46, 47, 54, 72; girls' average = 46

Do these figures support the theory that boys are better than girls in this subject?

Null hypothesis–that boys and girls are equally good at the subject.

From the sample of boys $\bar{x}_1 = 55 \qquad s_1^2 = 540.57$

From the sample of girls $\bar{x}_2 = 46 \qquad s_2^2 = 156.86$

Applying the F test to test that population variances are not different, gives .

$$F = \frac{540.57}{156.86} = 3.46 \text{ (not significant)} \qquad \upsilon_1 = 7 \qquad \upsilon_2 = 7$$

Best estimate of population variance by pooling

$$s^2 = \frac{(n_1-1)s_1^2 + (n_2-1)s_2^2}{n_1+n_2-2} = \frac{7 \times 540.57 + 7 \times 156.86}{(8+8-2)} \doteqdot 349.0$$

∴ Standard error of difference between means

$$\epsilon(\bar{x}_1 - \bar{x}_2) = \sqrt{[349(\tfrac{1}{8}+\tfrac{1}{8})]} = 9.35$$

$$\therefore \quad t = \frac{(55-46)}{9.35} - 0 = \frac{9}{9.35} = 0.96$$

with 14 degrees of freedom.

From table 7*

$$t_{0.025} = 2.14 \text{ (two sided 5\%)}$$

∴ There is no evidence from these data that boys are better than girls. (see discussion of example 5, chapter 7, p 153).

6. In designing a trial to test whether or not the conversion of a machine has reduced its variability, a sample of 20 on the new process is taken.

Previous machine standard deviation before conversion = 2.8 mm. For the new process, calculated from sample of 20, standard deviation = 1.7 mm.

What is the significance of this test?

It can be assumed that the process change could not have increased the variation of the process.

Null hypothesis—that no change has occurred in process variation. Thus, a one-tailed test can be used

$$F = \frac{2.8^2}{1.7^2} = 2.71 \qquad \upsilon_1 = \infty, \qquad \upsilon_2 = 19 \text{ (use 18)}$$

from table 9*. Thus

$$F_{0.05} = 1.92 \qquad F_{0.01} = 2.57$$

Therefore, the result is highly significant and the change can be assumed to have reduced the process variation.

8.3 Problems for Solution

1. Three women take an advanced typing course in order to increase their speed. Before the course their rates are 40, 42, 40 words per minute. After the course their speeds are, 45, 50 and 42 respectively. Is the course effective?

2. Table 8.6 gives the data obtained in an analysis of the labour turnover records of the departments of a factory. Is there any evidence that departmental factors affect labour turnover and if so, which departments?

Department	Average labour force	Number of leavers/year
A	60	15
B	184	16
C	162	15
D	56	12
E	30	4
F	166	25
G	182	25
H	204	18

Table 8.6

3. Table 8.7 gives the data obtained on process times of two types of machine. Is machine A more variable than machine B?

	Machine A	Machine B
Average time	2.5	2.3
Standard deviation	0.5	0.2
Sample size	100	80

Table 8.7

4. A change made to a process was tested by timing two sets of different workers. Those using the new process completed the job in

32, 32, 33, 33, 33, 34, 34, 35, 39, 45 s

Using the old process, another group completed it in

31, 32, 32, 33, 33, 34, 37, 43, 47, 48 s

Is the new process quicker?

5. In designing a trial to test whether or not the conversion of a machine has reduced its variability, a sample of 13 items is taken from the new process. Previous machine standard deviation before conversion = 2.8 mm; standard deviation from new process = 1.7 mm.

Is a reduced variability demonstrated?

6. The number of cars per hour passing an intersection, counted from 11 p.m. to 12 p.m. over nine days was 7, 10, 5, 1, 0, 6, 11, 4, 9.

Does this represent an increase over the previous average per hour of three cars?

7. In a time study, only 18 readings of an element could be taken as the order was nearly finished. They were as follows, in minutes

0.12, 0.14, 0.16, 0.12, 0.12, 0.17, 0.15, 0.14, 0.12,

0.11, 0.12, 0.12, 0.12, 0.15, 0.17, 0.13, 0.14, 0.14

Within what limits (95% confidence) would you expect the actual average time for this element to lie?

8. A coin is tossed 200 times and heads appear only 83 times. Is the coin biased?

9. A new advertising campaign is tried out in addition to normal advertising in six selected areas and the sales for a three-month period compared with

those of the six areas before the special campaign. The data are given in table 8.8. Has the new campaign had any effect on the sales?

	Sales before campaign	Sales after campaign
Area 1	£2000	£2500
2	£3600	£3000
3	£2500	£3100
4	£3000	£2800
5	£2800	£3400
6	£2900	£3200

Table 8.8

8.4 Solution to Problems

1. The null hypothesis is set up—that the advanced typing course will not affect the speed of typists.

This is a paired 't' test since by considering the differences only, the variation due to varying basic efficiency of the individuals is eliminated.

Typist	Difference x	x^2
1	5	25
2	8	64
3	2	4
	$\Sigma x_i = 15$	$\Sigma x_i^2 = 93$

$\bar{x} = 5$ Table 8.9

Estimated population variance

$$s^2 = \frac{\Sigma x_i^2 - \dfrac{(\Sigma x_i)^2}{n}}{n-1} = \frac{93 - \dfrac{(15)^2}{3}}{2} = 9 \quad \therefore \quad s = 3$$

Thus

$$t = \frac{\bar{x} - 0}{\dfrac{3}{\sqrt{3}}} = \frac{5 - 0}{\dfrac{3}{\sqrt{3}}} = 2.89$$

with 2 degrees of freedom.

Reference to table 7* (using two-tailed test)

$$t_{0.05/2} = 4.303 \qquad t_{0.010/2} = 9.925$$

or the result is not significant.

Thus there is no evidence from this sample that the new course improves speed.

It is not surprising that no evidence was found from this trial because of the small sample taken. In practice, when more girls were tested the new course was shown to be more effective, illustrating that the result *not significant* does not mean *no difference* but that *no evidence* of a difference has been found.

2. Null hypothesis: there is no difference in turnover rate between departments. The expected number of leavers and χ^2 contributions are given in table 8.10.

Dept.	Average labour force/year	Number of leavers/year (O_i)	Expected number of leavers/year (E_i)	Contribution to χ^2
A	60	15	7.5	7.5
B	184	16	23.0	2.1
C	162	15	20.0	1.2
D } E }	56 } 30 } 86	12 } 4 } 16	7.0 } 3.8 } 10.8	2.5
F	166	25	20.8	0.9
G	182	25	22.8	0.2
H	204	18	25.5	2.2
Totals	1044	130	130.4	$\chi^2 = 16.6$

Table 8.10

Since in department E, the expected number of leavers is less than five, it has to be grouped with another department. It is logical to group it with a similar department or one whose effect would be expected to be similar on number of leavers. Here, having no other *a priori* logic, since there is little difference between the observed and expected frequency for department E, it has little effect. Here it is combined with department D, the next smallest.

$$\therefore \quad \text{Average turnover rate} = \frac{130}{1044} \times 100\% = 12.5\% \text{ per year}$$

∴ Expected number of leavers per year in department A

$$= \frac{12.5}{100} \times 60 = 7.5$$

and so on.

Thus, $\chi^2 = 16.6$ with $(7-1)$ or 6 degrees of freedom since only the total was used to set up hypothesis.

Reference to table 8*

$$\chi^2_{0.05} = 12.592 \qquad \chi^2_{0.01} = 16.812 \qquad \chi^2_{0.001} = 22.457$$

Thus result is significant at 1/20 level, or there is evidence of differences between departments.

When such a result is obtained then it is usually possible to isolate the heterogeneous departments by locating the department with the largest contribution to χ^2. Providing the χ^2 is significant at 1 degree of freedom or exceeds 3.841 at 1/20 significance level, this department should be excluded from the data and the analysis repeated until χ^2 is not significant. If the result is significant with no single contribution greater than 3.841 then conclusion can only be drawn that heterogeneity is not due to one or two specific departments but general variations between all.

The results of repeating the analysis excluding department A are given in table 8.11.

Dept.	Average labour force/year	Number of leavers/year	Expected number of leavers/year	Contribution to χ^2
B	184	16	21.5	1.40
C	162	15	19.0	0.84
D	56 } 86	12 } 16	10.0	3.60
E	30	4		
F	166	25	19.4	1.61
G	182	25	21.3	0.64
H	204	18	23.8	1.41
Totals	984	115	115.0	9.50

Table 8.11

$$\text{Average turnover rate} = \frac{115}{984} \times 100\% = 11.7\%$$

Thus $\chi^2 = 9.50$ with $(6-1)$ or 5 degrees of freedom

$$\chi^2_{0.05} = 11.070 \qquad \chi^2_{0.01} = 15.086 \qquad \chi^2_{0.001} = 20.517$$

or result is not significant, there is no evidence of differences between remaining departments. It would however be worth checking further on department D since its contribution to total χ^2 has been 'watered-down' in having the data from E combined with it.

3. This is an 'F' test.
   Assume no *a priori* knowledge–use two-tailed test.

$$\therefore \quad F = \frac{0.5^2}{0.2^2} = 6.25 \qquad \upsilon_1 = 99 \qquad \upsilon_2 = 79$$

Referring to table 9* use $\upsilon_1 \doteqdot 24$, $\upsilon_2 \doteqdot 60$ (on safe side)

$$F_{0.05} = 1.88 \qquad F_{0.002} = 2.69$$

Clearly present value is highly *significant*, the product from machine A is more variable than the product from machine B.

4. Null hypothesis–the change to the process has *not* affected the time.
   Let x_1 = time of new process
   x_2 = time of old process

then, mean of new process $\bar{x}_1 = 35$ s.
Variance of new process estimated from sample

$$s_1^2 = \frac{\Sigma(x_i - \bar{x}_1)^2}{9} = 16.44$$

Mean of old process $\bar{x}_2 = 37$ s.
Variance of old process estimated from sample

$$s_2^2 = 42.67$$

In order to apply the 't' test, the variances of the two populations must be the same.

Using the 'F' test to test that population variances are the same, gives

$$F = \frac{42.67}{16.44} = 2.59 \quad \text{with} \quad \upsilon_1 = 9 \text{ degrees of freedom}$$

$$\upsilon_2 = 9 \text{ degrees of freedom}$$

Table 9* shows that this is not significant–or there is no evidence of difference in population variances.

Thus pooling the two estimates s_1^2 and s_2^2 to give best estimate

$$s^2 = \frac{(n_1 - 1)s_1^2 + (n_2 - 1)s_2^2}{n_1 + n_2 - 2} = 29.6$$

Standard error of the difference between the means

$$\epsilon(\bar{x}_1 - x_2) = \sqrt{[29.6\,(\tfrac{1}{10} + \tfrac{1}{10})]} = 2.43$$

$$\therefore \quad t = \frac{(37-35)-0}{2.43} = 0.82$$

with 18 degrees of freedom.

From table 7*

$$t_{0.05/2} = 2.101 \qquad t_{0.02/2} = 2.878$$

or result is not significant at 0.05 level—there is no evidence that the change has reduced the time.

5. Null hypothesis—conversion has *not* reduced the variability. Thus

$$F = \frac{2.8^2}{1.7^2} = 2.71 \qquad \upsilon_1 = \infty \text{ degrees of freedom}$$
$$\upsilon_2 = 12 \text{ degrees of freedom}$$

Referring to table 9* for $\upsilon_1 = \infty, \upsilon_2 = 12$ gives

$$F_{0.05} = 2.30 \qquad F_{0.01} = 3.36$$

The result is significant at the 5% level but not at the 1% level. Some further sampling would probably be in order so as to reduce the errors involved in reaching a decision.

Strictly, the one-sided F test (used because there is, say, prior knowledge that the conversion cannot possibly increase product variation but may reduce it) should be applied as follows.

Observed

$$F = \frac{1.7^2}{2.8^2} = 0.37$$

The lower 5% point of F with $\upsilon_1 = 12$ and $\upsilon_2 = \infty$ is obtained from table 9* as

$$F_{0.95,\,12,\,\infty} = \frac{1}{F_{0.05,\,\infty,\,12}} = \frac{1}{2.30} = 0.435$$

Since the observed value of F is lower than this, the reduction in variation is significant (statistically) at the 5% level.

The lower 1% point of F is

$$\frac{1}{3.36} = 0.30$$

and the observed F is not significantly low at this level.

6. Null hypothesis: there has been *no* increase in number of cars.

From sample

$$\Sigma x_i = 53 \qquad \Sigma x_i^2 = 429$$

$\therefore$ Sample mean

$$\bar{x} = 5.9$$

Variance

$$s^2 = \frac{429 - \dfrac{(53)^2}{9}}{8} = 14.6$$

Standard deviation

$$s = 3.82$$

$$\therefore \quad t = \frac{5.9 - 3}{\dfrac{3.82}{\sqrt{9}}} = 2.28$$

with 8 degrees of freedom.

From table 7*

$$t_{0.05/2} = 2.306$$

Thus the result is not quite significant at the 5% level. On the present data no real increase in mean traffic flow is shown.

7. Sample mean

$$\bar{x} = \frac{\Sigma x_i}{n} = 0.136 \text{ min}$$

Estimate of population variance

$$s^2 = \frac{\Sigma(x_i - \bar{x})^2}{n-1} = 0.000212 \qquad s = 0.0146 \text{ min}$$

Let μ_0 = unknown true population average. Then for 95% confidence

$$\bar{x} - t_{0.025}\frac{s}{\sqrt{n}} < \mu_0 < \bar{x} + t_{0.025}\frac{s}{\sqrt{n}}$$

$$0.136 - 2.11\frac{0.0146}{\sqrt{18}} < \mu_0 < 0.136 + 2.11\frac{0.0146}{\sqrt{18}}$$

or inside limits 0.136 ± 0.0073.

8. This problem will be solved using two alternative methods–the 'u' test and the χ^2 test.

1st Method–the 'u' Test

Hypothesis–the coin is unbiased.

$\therefore$ Probability of a head = 0.50

Sampling distribution of number of heads in 200 trials has

$$\mu = np = 200 \times 0.50 = 100$$

$$\sigma = \sqrt{[np(1-p)]} = \sqrt{200 \times 0.5 \times 0.5} = 7.07$$

$$\therefore \quad u = \frac{83.5-100}{7.07} = -\frac{16.5}{7.07} = -2.33$$

From table 3*, probability of 83 or fewer heads = 0.01; by symmetry the possibility of 117 or more heads is 0.01.

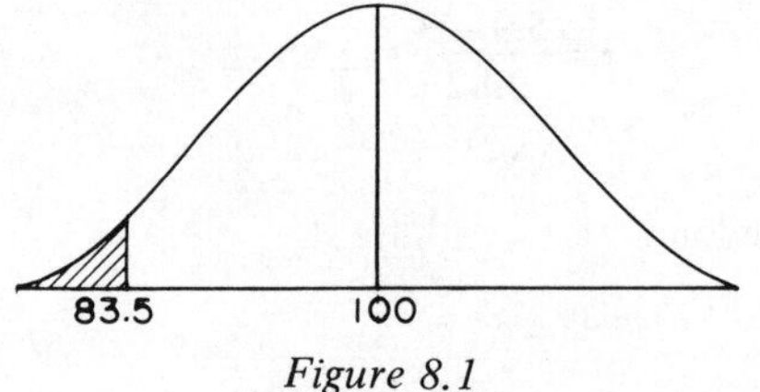

Figure 8.1

2nd Method–the 'χ^2' Test

		Heads	Tails
Observed	*O*	83	117
Expected	*E*	100	100

Table 8.12

	Heads	Tails
O	83.5	116.5
E	100	100

Table 8.13. Using Yate's correction

$$\therefore \quad \chi^2 = \frac{(83.5-100)^2}{100} + \frac{(116.5-100)^2}{100} = 5.445$$

with 1 degree of freedom.

From table 8* the probability of χ^2 this high, or higher, is approximately 0.02.

However, in calculating χ^2, the tabulation includes both tails of the normal distribution of which it is the sum of the squares.

Hence the probability of getting $\chi^2 > 5.445$ is 0.02 if both the probability of 83 or smaller and also of 117 or more are included.

Thus the probability of 83 or smaller = 0.01, which agrees with the result by the 'u' test.

9. Here, since from *a priori* knowledge, it can be stated that the new campaign can only increase the sales rate. Then a one-tailed test can be used for extra 'power' in the test.

Again the paired 't' test is applicable. Null hypothesis–new campaign has not increased the sales.

Area	Difference in sales x
1	+500
2	−600
3	+600
4	−200
5	+600
6	+300
Average	+200

Table 8.14

Code data

$$x' = \frac{x}{100}$$

Thus $\bar{x}' = 2$

$$s_{x'}^2 = \frac{\Sigma(x' - \bar{x}')^2}{5} = 24.4, \qquad s'_x = 4.93$$

$$\therefore \quad t = \frac{2 - 0}{\frac{4.93}{\sqrt{6}}} = 0.99$$

with 5 degrees of freedom

$$t_{0.05} = 2.015 \text{ (for one tail test)}$$

or result is not significant; there is no evidence of an increase in sales rate.

8.5 Practical Laboratory Experiments and Demonstrations

The concept of significance is perhaps one of the most difficult to grasp in statistics, i.e. that one cannot prove a hypothesis, only offer evidence on a probability basis for its rejection.

Here again practical participative laboratory experimentation gives the most effective vehicle for putting across this concept.

8.5.1 *Experiment 14—the 't' Test of Significance*

(This experiment is from the Laboratory Manual pages 62-65.)

Given to students in groups of two or three after lectures on significance testing and 't' test of means.

In this experiment, use is made of the two normal populations of rods supplied in the Kit.† While it is appreciated that realism can be introduced to experiments by using components from industry, experience has shown the necessity of having standard populations available, especially as they are used extensively throughout the experiments.

In Appendix 1 the instruction sheet, recording forms and analysis and summary sheets for the experiment are given together with a set of results obtained.

	Red rod population	Yellow rod population
Mean μ	6.0	6.2
Standard deviation σ	0.2	0.2

Table 8.15

The population parameters are given in table 8.15. These parameters are chosen so that for the first part of the experiment with sample sizes $n = 10$, approximately half the groups will establish a significant difference between the populations while the other half will show no significant difference at the 5% probability level. Since each group summarises the results of all the groups, this experiment brings out much more clearly than any lecture could do, the concept of significance.

In the second part of the experiment where each sample size is increased to 30, the probability is such that all groups generally establish (95% probability) a significant difference. The experiment demonstrates that there is a connection between the two types of error inherent in hypothesis testing by sampling and the amount of sampling carried out. To complete this experiment, including the full analysis, takes approximately 40 min.

† Available from Technical Prototypes, 1A Westholme Street, Leicester.

8.5.2 *Experiment 15–the 'F' Test*

(This experiment is described in pages 66-68 of the Laboratory Manual.)

The same rod populations as for experiment 14 again demonstrate the basic concepts of the test.

8.5.3 *Experiment 16–Estimation of Population Mean*

(Pages 69-71 of Laboratory Manual.)

8.5.4 *Experiment 17–Estimation of Population Mean (Small Sample)*

(Pages 72-74 of Laboratory Manual.)

8.5.5 *Experiment 18–Estimation of Population Variance*

(Pages 75-76 of Laboratory Manual.)

Note: All these experiments use the standard rod populations supplied with the Statistical Kit No. 1.

8.5.6 *Experiment 19–The* χ^2 *Test*

Using data from experiment 1 this experiment is described on pages 77-79 of the Laboratory Manual.

Appendix 1

Object

To test whether the means of two normal populations are significantly different and to demonstrate the effect of sample size on the result of the test.

Method

Take a random sample of size 10 from *each* of the two populations (red and yellow rods) and record the lengths in table 1. Return the rods to the appropriate population.

Also take a random sample of size 30 (a few rods at a time) from *each* of the two populations (red and yellow rods) and record the lengths in table 2.

Analysis

(1) Code the data, as indicated, in tables 1 and 2.
(2) Calculate the observed value of 't' for the two samples of size 10 and again for the samples of size 30.
(3) Summarise your results with those of other groups in tables 3 and 4. Observe whether a significant difference is obtained more often with the samples of size 30 than with the smaller samples.

Notes

The 't' test used is only valid provided the variances of the two populations are equal. This requirement is, in fact, satisfied in the present experiment.

Table I

Yellow population

Rod lengths	Coded data x'	x'^2
6·1	0·3	0·09
5·8	0·0	0·0
6·2	0·4	0·16
6·1	0·3	0·09
6·2	0·4	0·16
6·0	0·2	0·04
6·3	0·5	0·25
6·2	0·4	0·16
6·3	0·5	0·25
6·5	0·7	0·49
$\Sigma x =$	3·7	1·69 $= \Sigma x'^2$

Red population

Rod lengths	Coded data y'	y'^2
5·6	0·0	0·0
6·2	0·6	0·36
6·2	0·6	0·36
5·8	0·2	0·04
5·6	0·0	0·0
5·8	0·2	0·04
6·0	0·4	0·16
6·2	0·6	0·36
6·0	0·4	0·16
6·3	0·7	0·49
$\Sigma y' =$	3·7	1·97 $= \Sigma y'^2$

In order to reduce the subsequent arithmetic, and to keep all numbers positive, the coded values, x', are used in the calculation. The coded data can be obtained by subtracting from all readings, the smallest observed rod length in the sample. The coded values, y', may be obtained in a similar way for the sample of red rods.

If a is the length of the shortest yellow rod in the sample, the mean, $\bar{x}$, of the sample is

$$\bar{x} = a + \frac{\Sigma x'}{10} = 5.8 + \frac{3.7}{10} = 6.17$$

The variance, s_x^2, of the yellow sample is

$$s_x^2 = \frac{\Sigma x'^2 - \frac{(\Sigma x')^2}{10}}{9} = \frac{1.69 - \frac{3.7^2}{10}}{9} = \frac{0.32}{9} = 0.0355$$

If b is the length of the shortest red rod in the sample, the mean, $\bar{y}$, of the sample is

$$\bar{y} = b + \frac{\Sigma y'}{10} = 5.97$$

The variance, s_y^2, of the red sample is

$$s_y^2 = \frac{\Sigma y'^2 - \frac{(\Sigma y')^2}{10}}{9} = \frac{1.97 - 1.37}{9} = 0.0667$$

The pooled estimate of variance, s^2, is

$$s^2 = \frac{\Sigma x'^2 - \frac{(\Sigma x')^2}{18} + \Sigma y'^2 - \frac{(\Sigma y')^2}{10}}{18} = 0.0512$$

$$t = \frac{\bar{x} - \bar{y}}{s\sqrt{\frac{1}{n_x} + \frac{1}{n_y}}} = \frac{6.17 - 5.97}{\sqrt{0.0512}\sqrt{\frac{1}{10} + \frac{1}{10}}} = \frac{0.447}{\sqrt{0.0512}}$$

$$= 1.96$$

Table 2

Yellow population – (30)

Rod lengths	Coded data x'	x'^2
6·0	0·3	0·09
6·3	0·6	0·36
5·7	0·0	0·0
6·0	0·3	0·09
5·9	0·2	0·04
5·9	0·2	0·04
5·9	0·2	0·04
6·3	0·6	0·36
6·2	0·5	0·25
6·2	0·5	0·25
6·3	0·6	0·36
6·0	0·3	0·09
6·1	0·4	0·16
6·4	0·7	0·49
5·9	0·2	0·04
6·0	0·3	0·09
6·1	0·4	0·16
6·1	0·4	0·16
6·1	0·4	0·16
6·4	0·7	0·49
6·5	0·8	0·64
6·0	0·3	0·09
6·0	0·3	0·09
6·3	0·6	0·36
6·2	0·5	0·25
6·2	0·5	0·25
5·8	0·1	0·01
6·2	0·5	0·25
5·9	0·2	0·04
6·1	0·4	0·16
$\Sigma x'$ =	12·0	5·86 = $\Sigma x'^2$

Red population – (30)

Rod lengths	Coded data y'	y'^2
6·3	0·7	0·49
6·0	0·4	0·16
6·1	0·5	0·25
6·1	0·5	0·25
5·9	0·3	0·09
6·0	0·4	0·16
6·0	0·4	0·16
6·0	0·4	0·16
6·0	0·4	0·16
6·0	0·4	0·16
6·2	0·6	0·36
5·8	0·2	0·04
6·1	0·5	0·25
5·9	0·3	0·09
5·8	0·2	0·04
5·9	0·3	0·09
6·1	0·5	0·25
6·1	0·5	0·25
5·9	0·3	0·09
6·2	0·6	0·36
5·7	0·1	0·01
5·6	0·0	0·0
6·1	0·5	0·25
6·2	0·6	0·36
5·9	0·3	0·09
6·1	0·5	0·25
5·9	0·3	0·09
5·8	0·2	0·04
6·0	0·4	0·16
5·7	0·1	0·01
$\Sigma y'$ =	11·4	5·12 = $\Sigma y'^2$

The analysis is exactly similar to that for samples of size 10.

If a and b denote the lengths of the shortest yellow and red rods (in samples of 30), respectively

$$\bar{x} = a \div \frac{\Sigma x'}{30} = 6.10$$

$$\bar{x} = b + \frac{\Sigma y'}{30} = 5.98$$

The pooled estimate of variance, s^2, is

$$s^2 = \frac{\Sigma x'^2 - \frac{(\Sigma x')^2}{30} + \Sigma y'^2 - \frac{(\Sigma y')^2}{30}}{30 + 30 - 2}$$

$$t = \frac{(\bar{x} - \bar{y})}{a\sqrt{\frac{1}{n_x} + \frac{1}{n_y}}}$$

which reduces to

$$t = \frac{(\bar{x} - \bar{y})\sqrt{870}}{\sqrt{\Sigma x'^2 + \Sigma y'^2 - \frac{1}{30}[(\Sigma x')^2 + (\Sigma y')^2]}}$$

$$= \frac{12 \times 29.5}{\sqrt{10.98 - \frac{1}{30}[144 + 130]}}$$

$$= \frac{3.54}{1.36} = 2.6$$

Table 3

Summary Table — samples of size 10

Group	Sample means		Difference	Value of	Whether significant at 5% level (two-tail test)
	$\bar{x}$	$\bar{y}$	$(\bar{x}-\bar{y})$	t	
①	6·17	5·97	0·20	1·96	NO
2	6·11	6·01	0·10	1·27	NO
3	6·38	5·94	0·44	4·3	YES
4	6·25	6·02	0·23	2·16	YES
5	6·22	5·89	0·33	3·63	YES
6	6·09	5·96	0·13	1·53	NO
7	6·13	6·13	0	0	NO
8	6·17	5·97	0·20	2·44	YES

The value of $|t|$ which must be exceeded for the observed difference to be significant at the 5% level = 2·101

Table 4

Summary Table — samples of size 30

Group	Sample means		Difference	Value of	Whether significant at 5% level (two-tail test)
	$\bar{x}$	$\bar{y}$	$(\bar{x}-\bar{y})$	t	
①	6·10	5·98	0·12	2·6	YES
2	6·15	6·03	0·12	2·73	YES
3	6·29	5·97	0·32	5·88	YES
4	6·18	5·99	0·19	3·92	YES
5	6·18	5·96	0·22	4·06	YES
6	6·12	6·06	0·06	1·17	NO
7	6·19	5·99	0·20	3·64	YES
8	6·22	5·94	0·28	6·75	YES

The value of $|t|$ which must be exceeded for the observed difference to be significant at the 5% level = 2·002

9 Linear regression theory

9.1 Syllabus

Assumption for use of regression theory; least squares; standard errors; confidence limits; prediction limits; correlation coefficient and its meaning in regression analysis; transformations to give linear regression.

9.2 Résumé of Theory Covered

9.2.1 *Basic Concepts*

Regression analysis is concerned with the relationship between variables. In this chapter, only the linear relationship between a dependent variable, y, and an independent variable, x, will be discussed.

Regression analysis can be extended to cover curvilinear relationships between two variables and the relationship between a variable y and m other variables $x_1, x_2, \ldots x_m$. This is called multi-regression analysis and details can be found in textbooks on mathematical statistics.

The data for regression analysis may take two forms:

(1) The natural pairing of variables such as: height and weight, height of son and height of father, the output of a department per week and the average cost, or the sales of a product in an area and the advertising expenditure in that area.

(2) The independent variable x is given assigned values and for each value of x, a range of values of y is obtained. This type of data normally arises when the experimental design is under the control of the analyst and data in this form are from many points of view preferable to data of class (1). For example, in establishing relationships between cutting tool life and speed, the experimenter may vary speed (x) over a finite number of values and then take a number of observations of tool life (y) at each of these levels.

Note: It is important to appreciate clearly that the regression relationship calculated only holds over the range of variation of x used in the calculation. Any extrapolation of the relationships can only be carried out based upon *a priori* knowledge or assumptions that this relationship will hold for other values of x.

9.2.2 *Assumptions Required for Linear Regression Analysis*

The following assumptions are required for the use of regression theory and for the use of significance testing in the theory.

(1) The dependent variable (y) is normally distributed for each value of the independent variable (x).

(2) The independent variable x is either free from error or subject to negligible error only.

(3) The variance of y for all values of x is constant.

Note: It is also possible in advanced theory to apply regression analysis in cases where the variance of y is a function of x.

9.2.3 *Basic Theory*

The regression line is fitted by the method of least squares. Given the population theoretical regression line as

$$\eta = \alpha + \beta(x - \bar{x})$$

then the best estimate of this line is given by

$$Y = a + b(x - \bar{x})$$

where

$$a = \frac{\sum_i^n f_i y_i}{\sum_i^n f_i}$$

and

$$b = \frac{\Sigma f_i(y_i - \bar{y})(x_i - \bar{x})}{\Sigma f_i(x_i - \bar{x})^2}$$

a and b are unbiased estimates of α and β respectively.

These estimates minimise the residual variance of y about the regression line and for this reason the approach is known as the 'method of least squares'.

Note: In the first form of data where there is a natural pairing of the points, $f_i = 1$ for all i, and the regression coefficients are given by the following formulae

$$a = \frac{\sum_i^n y_i}{n} \simeq \alpha \qquad b = \frac{\sum_i^n (y_i - \bar{y})(x_i - \bar{x})}{\Sigma(x_i - \bar{x})^2} \simeq \beta$$

where n = number of pairs of observations.

Since this book is concerned with giving an introduction to the theory, the examples given will be for this case of paired variables. It should be stressed, however, that for cases where $f_i > 1$ a more rigorous theory can be developed and, in fact, a test for linearity can be incorporated into the analysis. Details of this more advanced analysis can be found in most mathematical statistics textbooks.

This omission of an independent test of linearity requires usually an *a priori* knowledge of linearity and this should in all cases be examined by drawing a scatter diagram.

9.2.4 *Significance Testing*

It is, of course, necessary not only to calculate the statistics '*a*' and '*b*' but also to be able to test their significance. This point cannot be stressed strongly enough. In addition it should be noted that even if a regression coefficient is found to be significant, it does not necessarily imply a causal relationship between the variables.

The standard errors of the coefficients are: standard error of a

$$\epsilon_a = \frac{\hat{s}}{\sqrt{n}}$$

standard error of b

$$\epsilon_b = \frac{\hat{s}}{\sqrt{\Sigma(x_i - \bar{x})^2}}$$

where s' = residual variance about regression line

$$= \frac{\Sigma(y_i - Y_i)^2}{n - 2}$$

where Y_i = estimate from regression line.

The significance of a and b can, therefore, be tested by the 't' test (see chapter 8) or, alternatively, as shown in some textbooks, by an 'F' test, (see for example Weatherburn, *A First Course in Mathematical Statistics*, C.U.P., pages 193 and 224, example 8).

The t-*Test of the Significance of an Observed Regression Coefficient* b

Set up the null hypothesis that $\beta = 0$, i.e. that there is no linear relationship between y and x and thus the values of y are independent of the values of x.

Remember that in this simple theory, it is necessary to assume that the only possible relation between y and x is a linear one.

Under the assumptions given in section 9.2.2, the statistic

$$\frac{b - \beta}{\epsilon_b}$$

will be distributed like Student's t with $(n-2)$ degrees of freedom. The degrees of freedom of ϵ_b are 2 less than the number of points (pairs of values of x and y) since the residual sum of squares about the fitted regression line is subject to two independent constraints corresponding to the two constants calculated from the data and used to fit the regression line.

The value of

$$t = \frac{b-0}{\epsilon_b}$$

given by the data can be referred to table 7* of the Statistical Tables and if it is significantly large, judged usually on a two-sided basis, there is thus evidence of a linear relationship between y and x.

9.2.5 *Confidence Limits for the Regression Line*

The standard error, ϵ_{Y_i}, of the regression estimate, Y_i, is given by

$$\sqrt{[\epsilon_a^2 + \epsilon_b^2 (x_i - \bar{x})^2]} = \sqrt{\left[\frac{s^2}{n} + \frac{s^2 (x_i - \bar{x})^2}{\sum\limits_i (x_i - \bar{x})^2}\right]}$$

$100(1-\alpha)\%$ confidence limits for the precision of estimation of the regression line are then given by $Y_i \pm t_{\alpha/2,\upsilon}\, \epsilon_{Y_i}$ for given x_i where $\upsilon = n-2$.

Note: The confidence limits are closest together at the average value $\bar{x}$, of the independent variable.

9.2.6 *Prediction Limits*

The confidence limits defined in section 9.2.5 relate to the position of the assumed 'true' regression line. If the relation is to be used to predict the value of y that would be observed corresponding to a given value of x, then, in addition to the uncertainty about the 'true' regression line, the scatter of individual values of y about this 'true' line must also be allowed for.

The standard error of a single value of y corresponding to a given value, x_i, is

$$\epsilon_{y_i} = \sqrt{\left[s + \frac{s^2}{n} + s^2 \frac{(x_i - \bar{x})^2}{\sum\limits_i (x_i - \bar{x})^2}\right]}$$

obtained by adding on the variance of a single value of y to the variance of the regression estimate, Y_i.

Thus, for a particular x_i, there is a probability of $100(1-\alpha)\%$ that the corresponding value of y that would be observed will lie in the interval

$$Y_i \pm t_{\alpha/2,\upsilon}\, \epsilon_{yi}$$

9.2.7 *Correlation Coefficient (r)*

A measure closely related to the regression coefficient (b) is the correlation coefficient (r).

The correlation coefficient (r) is a measure of the degree of (linear) association between the two variables and is defined as

$$r = \frac{\Sigma(x-\bar{x})(y-\bar{y})}{\sqrt{[\Sigma(x-\bar{x})^2\ \Sigma(y-\bar{y})^2]}}$$

The observed correlation coefficient can be tested for significant departure from zero but, as in the case of the regression coefficient, b, a significant value does not necessarily imply any causal relationship between x and y.

The residual variance about the regression line defined in section 9.2.4 as the sum of the squared deviations of each observed value of y from its estimated value using the fitted regression equation, this sum being divided by $(n-2)$ its degrees of freedom, is related to the correlation coefficient and to the total variance of y.

Thus

$$s^2 = \frac{\sum_i (y_i - Y_i)^2}{n-2} = (1-r^2)\frac{\sum_i (y_i-\bar{y})^2}{(n-2)} = (1-r^2)\frac{\sum_i (y_i-\bar{y})^2}{(n-1)}\ \frac{(n-1)}{(n-2)}$$

$$= s_y^2(1-r^2)\frac{(n-1)}{(n-2)}$$

which for large n is approximately equal to $s_y^2(1-r^2)$.

For large n, it follows that a useful interpretation of this result is that r^2 measures the proportion of the total variance of y that is 'explained' by the linear relation between y and x.

r^2 can take values between 0 and 1 inclusive and hence for any set of data, r will be in the range

$$-1 \leqslant r \leqslant +1$$

When $r = \pm 1$, then the total variance of y is completely explained by the variation in x or in other words the relationship is deterministic.

Figure 9.1 shows three sets of data with different values of the correlation coefficient (r). In the first two cases, the regression coefficient b is the same.

A further useful relationship is that between the regression coefficient (b) and the correlation coefficient (r) and which is

$$b = r \times \frac{s_y}{s_x}$$

where s_x^2 is the variance of the values of x, i.e.

$$\frac{\sum_i (x_i - \bar{x})^2}{n-1}$$

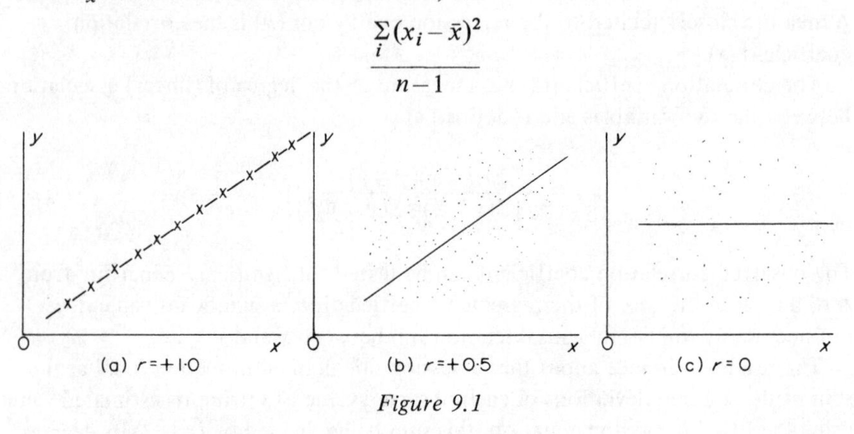

Figure 9.1

In practice, it is usual for all these calculations to be carried out on some form of calculating machine or computer, though there is no reason, apart from the tedious arithmetic involved, why they should not be done 'by hand' preferably with suitable coding of the data.

The coefficients are computed as follows

$$a = \frac{\sum_i^n y_i}{n}$$

$$b = \frac{\Sigma x_i y_i - \dfrac{(\Sigma x_i)(\Sigma y_i)}{n}}{\Sigma x_i^2 - \dfrac{(\Sigma x_i)^2}{n}}$$

The correlation coefficient

$$r = \frac{\Sigma x_i y_i - \dfrac{(\Sigma x_i)(\Sigma y_i)}{n}}{\sqrt{\left\{\left[\Sigma x_i^2 - \dfrac{(\Sigma x_i)^2}{n}\right]\left[\Sigma y_i^2 - \dfrac{(\Sigma y_i)^2}{n}\right]\right\}}}$$

Also the regression coefficient

$$b = r \times \frac{s_y}{s_x}$$

Thus the following totals are required for the computation

$$n, \quad \Sigma x_i, \quad \Sigma x_i^2, \quad \Sigma x_i y_i, \quad \Sigma y_i, \text{ and } \Sigma y_i^2.$$

9.2.8 *Transformations*

In some problems the relationship between the variables, when plotted or from *a priori* knowledge, is found not to be linear. In many of these cases it is possible to transform the variables to make use of linear regression theory.

For example, in his book *Statistical Theory with Engineering Applications* (Wiley), Hald discusses the problem of the relationship between tensile strength of cement (y) and its curing time (x).

From *a priori* knowledge a relationship of the form $y = A\mathrm{e}^{-B/x}$ is to be expected.

The simple logarithmic transformation therefore gives

$$\log_{10} y = \log_{10} A - \frac{B}{x} \log_{10} \mathrm{e}$$

or the logarithm of the tensile strength is a linear function of the reciprocal value of the curing time and the theory of linear regression can then be applied.

Note: The requirement that the variance of y is constant for all x must, of course, hold in the transformation and this must be checked. Usually a visual check is adequate.

9.2.9 *Example on the Use of Regression Theory*

The following example has been selected to illustrate the various concepts, computational methods and analysis.

In order to keep the computation inside reasonable limits, the number of observations has been kept small; in practice, however, in many actual problems hundreds of readings are involved, but with the use of computers the computation is no problem.

The data given in table 9.1 show the relationship between the scoring of post-graduate students in a numeracy test on interview and their performance in the final quantitative examination.

Student	1	2	3	4	5	6	7	8	9	10
Numeracy test score (pts)	200	175	385	300	350	125	440	315	275	230
Final exam performance (%)	55	45	71	61	62	50	74	67	65	52

Table 9.1

What is the best relationship between test score and final performance?

Before any analysis is started the scatter diagram must be plotted to test the assumption that the relationship is linear. This diagram shows no evidence of non-linearity (figure 9.2).

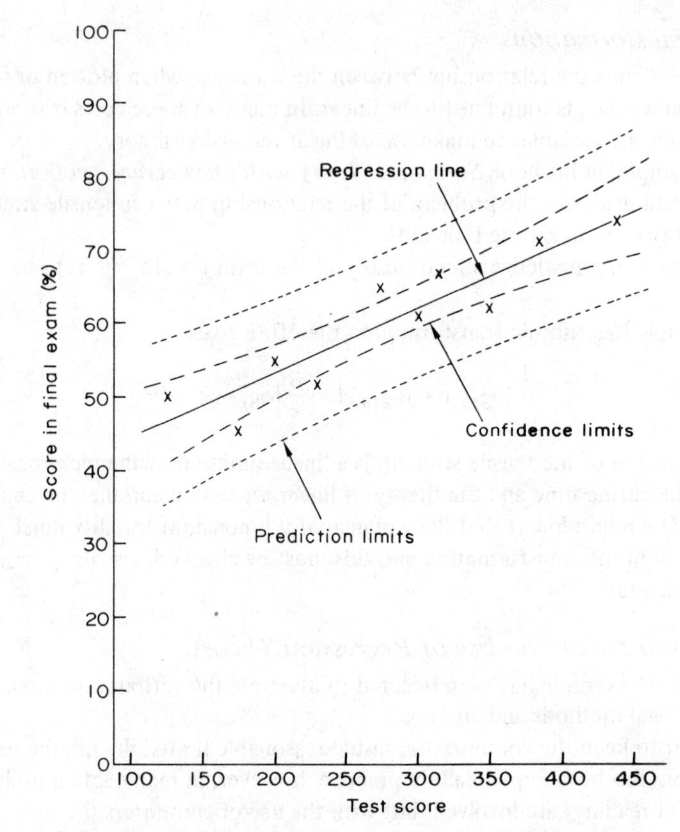

Figure 9.2. Regression line with 95% confidence limits and prediction limits.

Here y = final exam performance
x = numeracy test score.

No attempt has been made to code the data and the various summations required for analysis are given below

$$n = 10 \qquad \Sigma x_i y_i = 175\,960$$

$$\Sigma y_i = 602 \qquad \Sigma y_i^2 = 37\,050$$

$$\Sigma x_i = 2795 \qquad \Sigma x_i^2 = 868\,325$$

Total variance of x

$$s_x^2 = \frac{\Sigma x_i^2 - \dfrac{(\Sigma x)^2}{n}}{n-1} = \frac{868\,325 - \dfrac{(2795)^2}{10}}{9} = \frac{87\,122}{9} = 9680.3$$

Total variance of y

$$s_y^2 = \frac{37\,050 - \dfrac{(602)^2}{10}}{9} = \frac{809.6}{9} = 90.0$$

Correlation Coefficient

$$r = \sqrt{\left(\frac{175\,960 - \dfrac{2795 \times 602}{10}}{87\,122 \times 810}\right)} = \frac{7701}{8401} = 0.92$$

Regression Coefficients

$$a = \frac{\Sigma y_i}{n} = \frac{602}{10} = 60.2 \qquad b = r \times \frac{s_y}{s_x} = \frac{7701}{8401}\sqrt{\left(\frac{90}{9680}\right)} = 0.088$$

The regression line is thus given by

$$Y = 60.2 + 0.088\,(x - 279.5) = 35.6 + 0.088\,x$$

Residual Variance about the Regression Line

The approximate residual variance using the relation given in 9.2.7 is

$$s = s_y^2(1 - r^2) = 90(1 - 0.92^2) = 13.8$$

Note: This will be slightly in error through squaring a rounded value of r. A better approach, to ensure arithmetical accuracy, would be to calculate s^2 as

$$90.0\left(1 - \frac{7701^2}{8401^2}\right) = 14.4$$

However, in addition, in this example, n (=10) is not very large and the approximation used will lead to an underestimate of the actual residual variance. Using the exact expression gives

$$s^2 = \frac{s_y^2(1 - r^2)(n-1)}{(n-2)} = 14.4 \times \tfrac{9}{8} = 16.2$$

and this will be used in the remaining calculations since without this correction, the bias of the estimator is $-\frac{1}{9}$ (11%) of the true value.

The residual standard deviation, s, is $\sqrt{16.2} = 4.02$

Standard Errors of the Regression Coefficients

$$\epsilon_a = \frac{s}{\sqrt{n}} = \sqrt{\frac{s^2}{n}} = \sqrt{\left(\frac{16.2}{10}\right)} = 1.27$$

$$\epsilon_b = \frac{s}{\sqrt{[\Sigma(x-\bar{x})^2]}} = \sqrt{\left[\frac{s^2}{\Sigma(x-\bar{x})^2}\right]} = \sqrt{\left(\frac{16.2}{87\,122}\right)} = 0.0136$$

Significance of b
Assuming $E[b] = \beta = 0$, the observed value of t is

$$t = \frac{0.088 - 0}{0.0136} = 6.47$$

with 8 degrees of freedom.

Reference to table 7* shows that this value exceeds the 0.1% level of t (5.041 for the two-sided test) and hence the observed value of $b = +0.088$ is very significantly different from zero. This implies that there is a strong linear relation between y and x, i.e. between final quantitative examination performance and initial numeracy test score.

Confidence Limits for the Regression Line
The standard error, ϵ_{Y_i}, of the regression estimate is

$$\epsilon_{Y_i} = \sqrt{[\epsilon_a^2 + \epsilon_b^2(x_i - \bar{x})^2]} = \sqrt{[1.27^2 + 0.0136^2(x_i - 279.5)^2]}$$

Thus for

$$x_i = \bar{x} = 279.5, \qquad \epsilon_{Y_i} = \sqrt{1.27^2} = 1.27$$

$$x_i = 380 \text{ (or 179)}, \qquad \epsilon_{Y_i} = \sqrt{(1.27^2 + 0.0136^2 \times 100.5^2)} = 1.87$$

$$x_i = 440 \text{ (or 119)}, \qquad \epsilon_{Y_i} = \sqrt{(1.27^2 + 0.0136^2 \times 160.5^2)} = 2.53$$

For any given value of x_i, the confidence limits for the regression estimate (i.e. of the mean value of y for that value of x) are found as

$$Y_i \pm t_{\alpha/2,(n-2)}\, \epsilon_{Y_i}$$

For 95% limits, the appropriate value of t (table 7*) is 2.306; table 9.2 shows the derivation of the actual limits for a range of values x.

The scatter diagram (drawn before any computations were carried out, in order to check that the basic regression assumptions were not obviously violated), the fitted regression line and 95% confidence limits are shown in figure 9.2.

From figure 9.2 or table 9.2, there is 95% confidence that the *average* final examination percentage for all candidates who score 330 points in their initial numeracy test will lie between 61.3% and 67.9%.

x_i	Y_i	ϵ_{Y_i}	$2.31\ \epsilon_{Y_i}$	Lower 95% limit $(Y_i - 2.31\ \epsilon_{Y_i})$	Upper 95% limit $(Y_i + 2.31\ \epsilon_{Y_i})$
119	46.1	2.53	5.8	40.3	51.9
179	51.4	1.87	4.3	47.1	55.7
229	55.8	1.44	3.3	52.5	59.1
279.5	60.2	1.27	2.9	57.3	63.1
330	64.6	1.44	3.3	61.3	67.9
380	69.0	1.87	4.3	64.7	73.3
440	74.3	2.53	5.8	68.5	80.1

Table 9.2

Prediction Limits for a Single Value of y for Given x

The standard error, ϵ_{yi}, of a single value of y corresponding to a given value of x_i is

$$\epsilon_{yi} = \sqrt{[s^2 + \epsilon_a^2 + \epsilon_b^2(x_i - \bar{x})^2]}$$

Limits within which 95% of all possible values of y for a given x_i will lie are found as

$$Y_i \pm 2.31\ \epsilon_{yi}$$

These limits are calculated in table 9.3 and are also drawn in figure 9.2.

x_i	Y_i	ϵ_{yi}	Lower 95% prediction limit $(Y_i - 2.31\ \epsilon_{yi})$	Upper 95% prediction limit $(Y_i + 2.31\ \epsilon_{yi})$
119	46.1	4.8	35.0	57.2
179	51.4	4.4	41.2	61.6
229	55.8	4.3	45.9	65.7
279.5	60.2	4.2	50.5	69.9
330	64.6	4.3	54.7	74.5
380	69.0	4.4	58.8	79.2
440	74.3	4.8	63.2	85.4

Table 9.3

From the figures in table 9.2, it can be expected, for example, that 95% of candidates scoring 330 points in their numeracy test will achieve a final examination mark between 55% and 74% inclusive, 5% of candidates gaining marks outside this range.

Note: Such predictions are only likely to be at all valid if the sampled data used to calculate the regression relation are representative of the same population of students (and examination standards) for which the prediction is being made. In other words, care must be taken to see that inferences really do apply to the population or conditions for which they are made.

The danger of extrapolation has been mentioned. The regression equation indicates that students scoring zero in the test, on average, gain a final mark of 35.6%. This may be so but it is very likely that the relation between the two examination performances is not linear over all values of x. Conclusions on the given data should only be made for x in the range 125 to 440.

9.3 Problems for Solution

1. The shear strength of electric welds in metal sheets of various thickness is given in table 9.4.

Thickness of sheets (mm)	Shear strength of sheets (kg)
0.2	102
0.3	129
0.4	201
0.5	342
0.6	420
0.7	591
0.8	694
0.9	825
1.0	1014
1.1	1143
1.2	1219

Table 9.4

Calculate the linear relationship between strength and thickness and give the limits of accuracy of the regression line.

2. The following problem is based on an example in Ezekiel's *Methods of Correlation Analysis* and shows for 20 farms, the annual income in dollars together with the size of the farm in hectares (i.e. units of 10 000 m^2). The data are given in table 9.5.

Find the best linear relationship between the size of farm and income and

Size of farm (ha) (x)	Income ($) (y)
60	960
220	830
180	1260
80	610
120	590
100	900
170	820
110	880
160	860
230	760
70	1020
120	1080
240	960
160	700
90	800
110	1130
220	760
110	740
160	980
80	800

Table 9.5

state the limits of error in using this relationship to predict farm income from farm size.

3. The data obtained from a controlled experiment to determine the relationship between y and x are given below

x	5	10	15	20	30	40	55	65	80
y	7.2	14.7	21.0	27.5	30.0	35.0	37.3	40.2	41.8

Calculate the linear regression line.

4. A manufacturer of optical equipment has the following data on the unit cost of certain custom-made lenses and the number of units in each order.

Number of units	1	3	5	10	12	(x)
Cost per unit (£)	58	55	40	37	22	(y)

(a) Calculate the regression coefficients and thus the regression equation which will enable the manufacturer to predict the unit cost of these lenses in terms of the number of lenses contained in each order.

(b) Estimate the unit cost of an order for eight lenses.

5. The work of wrapping parcels of similar boxes was broken down into eight elements. The sum of the basic seconds per parcel (i.e. of these eight elements) together with the number of boxes in each parcel is given in table 9.5.

Number of boxes in parcel (x)	Sum of basic seconds per parcel (y)	Number of boxes in parcel (x)	Sum of basic seconds per parcel (y)
1	130	22	260
6	200	27	190
13	150	34	290
19	200	42	270

Table 9.5

(a) Calculate the constant basic seconds per parcel and the basic seconds for each additional box in the parcel.

Calculate the linear regression and test its significance.

(b) What would be the best estimate of the basic seconds for wrapping a parcel of 18 boxes?

6. A manufacturer of farm tools wishes to study the relationship between his sales and the income of farmers in a certain area. A sample of 11 regions showing the income level of farmers in that area, together with the total sales to the area, gave the data in table 9.6. Of what use is this information to the manufacturer?

Income level of farms in area ($)	Total sales to farms in area ($)	Income level of farms in area ($)	Total sales to farms in area ($)
1300	2800	1300	3000
900	1900	1200	2600
1400	3200	800	3300
1000	2400	1400	1500
800	1700	700	1600
900	2000		

Table 9.6

7. The following example illustrates the application of regression analysis to time series.

The annual sales of a product over eight years are given below

1960	1961	1962	1963	1964	1965	1966	1967
300	215	450	325	375	300	375	400

Estimate the best linear time trend and calculate confidence limits for forecasting.

9.4 Solutions to Problems

1. Let x = thickness of sheet (mm)

y = shear strength of sheet (kg)

$$n = 11 \qquad \Sigma x_i^2 = 6.49$$

$$\Sigma x_i = 7.7 \qquad \Sigma y_i^2 = 5\,692\,958$$

$$\Sigma y_i = 6680.0 \qquad \bar{x} = 0.7$$

$$\Sigma x_i y_i = 6008 \qquad \bar{y} = 607.3$$

Variance of x

$$s_x^2 = \frac{6.49 - \dfrac{(7.7)^2}{11}}{10} = \frac{1.10}{10} = 0.110$$

Total Variance of y

$$s_y^2 = \frac{5\,692\,958 - \dfrac{(6680)^2}{11}}{10} = \frac{1\,636\,376.2}{10} = 163\,637.6$$

Correlation Coefficient

$$r = \frac{6008 - \dfrac{(7.7)(6680)}{11}}{\sqrt{(1.10 \times 1\,636\,376.2)}} = +\,0.9928$$

The proportion of the total variance of y 'explained' by the linear regression relation between y and x is approximately 0.9928^2 or 98.6%.

Regression Line

$$a = \bar{y} = 607.3$$

$$b = r \times \frac{s_y}{s_x} = 0.9928 \times \sqrt{\frac{163\,637.6}{0.110}} = 1210.9$$

∴ The linear regression line is given as

$$Y - 607.3 = 1210.9\,(x - 0.7) \quad \text{or} \quad Y = 240.3 + 1210.9$$

Standard Errors

The estimated residual variance about the regression line is

$$s^2 = s_y^2(1 - r^2)\left(\frac{n-1}{n-2}\right) = 163\,637.6\,(1 - 0.9928^2)\,\frac{10}{9} = 2608.7$$

thus

$$\epsilon_a = \sqrt{\left(\frac{2608.7}{11}\right)} = 15.40 \qquad \epsilon_b = \sqrt{\left(\frac{2608.7}{1.10}\right)} = 48.7$$

Test of Significance of b

From the evidence of the scatter diagram and the high value of r, the observed value of b is expected to be significant. In confirmation, the test gives

$$t = \frac{1210.9 - 0}{48.7} = 24.9$$

a very highly significant value of t for 9 degrees of freedom.

Confidence Limits and Prediction Limits

The estimated standard error of the regression line is

$$\epsilon_{Y_i} = \sqrt{[\epsilon_a^2 + \epsilon_b^2(x - \bar{x})^2]}$$

and the estimated standard error of a single predicted value of y for given x is

$$\epsilon_{y_i} = \sqrt{[s^2 + \epsilon_a^2 + \epsilon_b^2(x - \bar{x})^2]}$$

Table 9.7 shows some values of these two standard errors for particular values of x, together with the 95% confidence and prediction limits using the appropriate t-value of 2.26 (9 degrees of freedom).

The information in this table, as well as the observed data are plotted in figure 9.3.

Notice that the fitted 'best' line does not go through the origin. In fact the origin is not contained within the 95% confidence interval for the 'true' regression line—which is equivalent to saying that the intercept of the fitted line is significantly (5% level) different from zero. From inspection of the observed data, there is a suggestion that the true relation curves towards the origin for low values of sheet thickness. In short, do not extrapolate for thickness values below 0.2 mm and bear in mind that the calculated relationship for sheet thicknesses of 0.2 mm and just above may underestimate the average shear strength of welds.

x_i	Y_i	95% confidence limits for regression line		95% prediction limits for single values	
		ϵ_{Yi}	$Y_i \pm 2.26\ \epsilon_{Yi}$	ϵ_{yi}	$Y_i \pm 2.26\ \epsilon_{yi}$
0.2	1.9	28.81	−63, 67	58.64	−131, 134
0.3	123.0	24.83	67, 179	56.79	−5, 251
0.4	244.1	21.23	196, 292	55.31	119, 369
0.5	365.2	18.22	324, 406	54.23	243, 488
0.6	486.2	16.15	450, 523	53.57	365, 607
0.7	607.3	15.40	572, 642	53.35	487, 728
0.8	728.4	16.15	692, 765	53.57	607, 849
0.9	849.5	18.22	808, 891	54.23	727, 972
1.0	970.6	21.23	923, 1019	55.31	846, 1096
1.1	1091.7	24.83	1036, 1148	56.79	963, 1220
1.2	1212.8	28.81	1148, 1278	58.64	1080, 1345

Table 9.7

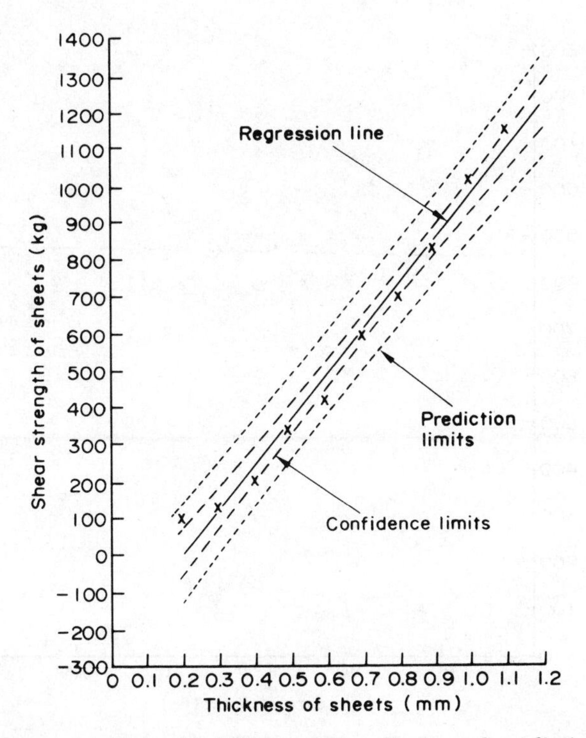

Figure 9.3. Regression line with 95% confidence limits and prediction limits.

In the following solutions, since the calculations are all similar to that of problem 1, the detailed computations are not given.

2. Here the scatter diagram (figure 9.4) shows little evidence of a relationship but, on the other hand, it does not offer any evidence against the *linearity* assumption so the computation is as follows.

$n = 20$ $\bar{x} = 139.5$

$\Sigma x = 2790$ $\bar{y} = 872.0$

$\Sigma y = 17\,440$ $s_x^2 = 3194.47$

$\Sigma x^2 = 449\,900$ $s_y^2 = 28711.58$

$\Sigma y^2 = 15\,753\,200$ $r = +0.0078$

$\Sigma xy = 2\,434\,300$ $s_x = 56.5$

$s_y = 169.4$

$$b = 0.0078 \times \frac{169.4}{56.5} = +0.02339$$

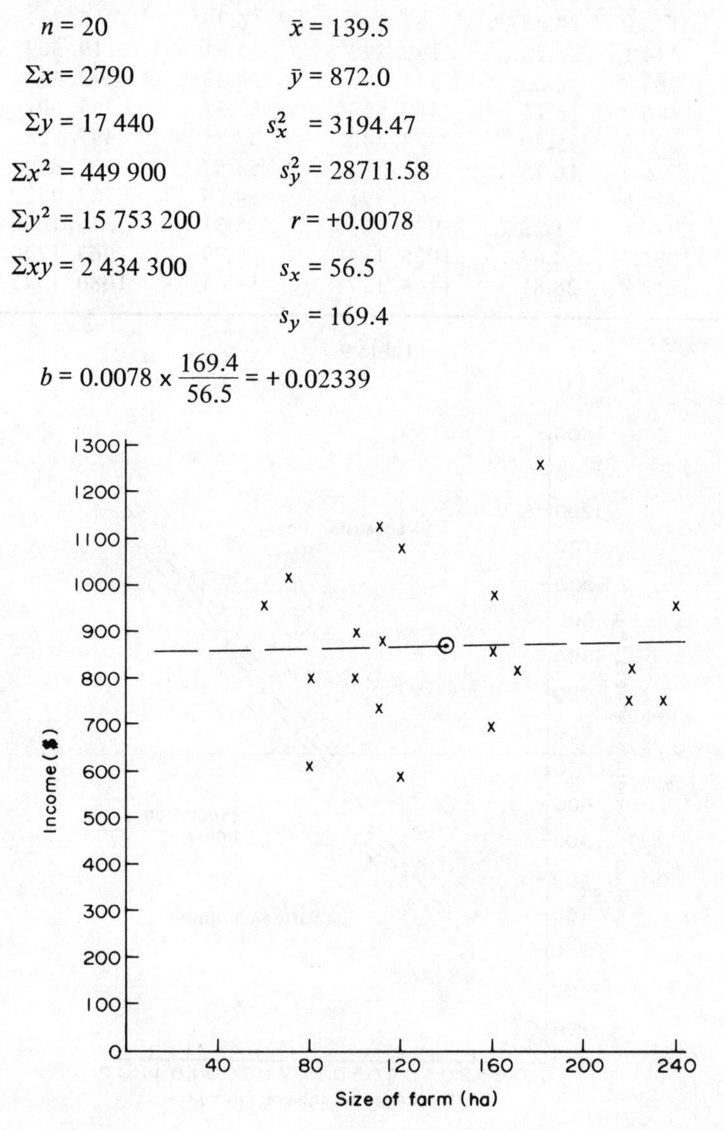

Figure 9.4

Regression Line

$$Y - 872 = 0.0234\,(x - 139.5) \qquad Y = 868.7 + 0.0234x$$

Significance of b

From inspection of the scatter diagram (figure 9.4) and the low value of r (the significance of which can be tested using table 10*), the observed value of b is not expected to differ significantly from zero.

Residual variance

$$s^2 = 28\,711.58(1 - 0.0078^2) \times \tfrac{19}{18} = 30\,305$$

Standard error of b

$$\epsilon_b = \sqrt{\left[\frac{30\,305}{449\,900 - \dfrac{(2790)^2}{20}}\right]} = \sqrt{\left(\frac{30\,305}{60\,695}\right)} = 0.71$$

Thus the observed value of

$$t = \frac{0.0234 - 0}{0.71} = 0.033$$

which is clearly not significant. (For the slope of the fitted regression line to be significantly different from zero, at the 5% level, the observed value of t would have to be numerically larger than 2.101.)

Thus, until further evidence to the contrary is obtained, farm income can be assumed to be independent of farm size, at least for the population of farms covered by the sample of 20 farms.

Since the data show no evidence of a relation between farm size and income, there is little point in retaining the fitted regression equation. The best estimate of the mean income of farms in the given population is therefore \$872.

Ninety-five per cent confidence limits for this *mean* income are given by

$$872 \pm 2.101 \times \frac{169.4}{\sqrt{20}} = 872 \pm 79.6 = \$792.4 \text{ to } \$951.6$$

Ninety-five per cent prediction limits for the income of an individual farm are given as

$$872 \pm 2.101 \times 169.4\sqrt{(1 + \tfrac{1}{20})} = \$872 \pm 364.7$$

$$= \$507.3 \text{ to } \$1236.7$$

3. This problem is of interest since the assumption of linearity can be quite safely rejected after drawing the scatter diagram (figure 9.5). There is therefore no point in trying to fit a single linear relationship to the data.

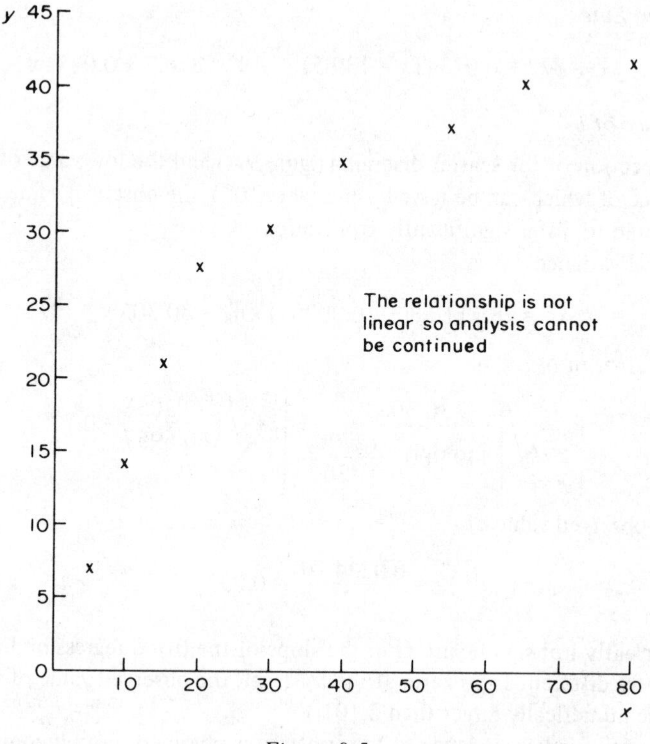

Figure 9.5

In practice either a polynomial or other mathematical function would be fitted to the observed data or else a suitable transformation of the values of either x or y or both would be used to give an approximately linear relation. In this latter case, the standard methods could be used to find the linear regression relation between the *transformed* y-values and the *transformed* x-values.

However, since both methods are beyond the scope of this chapter, the answer here is that linear regression analysis cannot validly be used directly with these data. Although there appears to be a relationship, it is not linear.

4. The scatter diagram (figure 9.6) indicates quite a strong relationship between unit cost and order size, and a simple linear relation would probably be adequate, at least in the range of order size considered. Such a simple model would be inadequate for extrapolation purposes since the cost per unit would be expected to tend towards a fixed minimum value as order size was increased indefinitely and therefore some sort of exponential relation would be a better fit for such purposes.

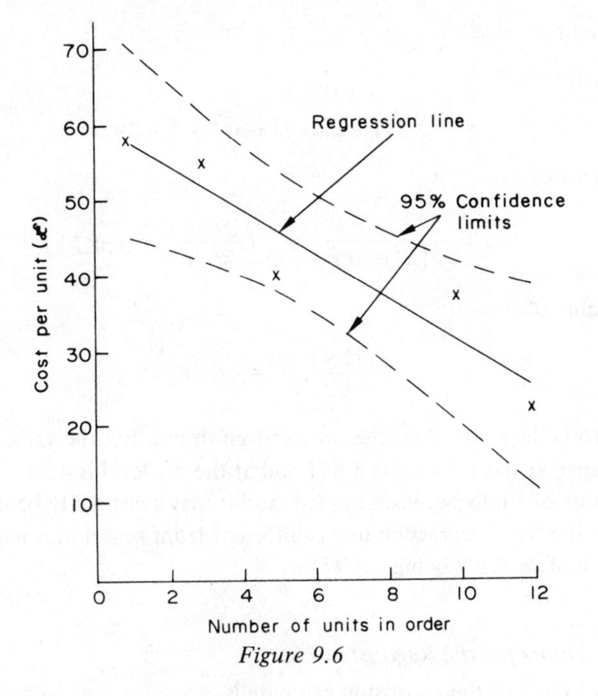

Figure 9.6

(a) The required totals of the basic data are

x = number of units in an order
y = cost per unit (£)

$$n = 5$$

$$\Sigma x = 31 \qquad \bar{x} = 6.2$$

$$\Sigma x^2 = 279 \qquad s_x^2 = 21.7$$

$$\Sigma y = 212 \qquad \bar{y} = 42.4$$

$$\Sigma y^2 = 9842 \qquad s_y^2 = 213.3$$

$$\Sigma xy = 1057 \qquad r = -0.9459$$

Regression Line

$$a = \bar{y} = 42.4$$

$$b = -0.9459 \times \sqrt{\left(\frac{213.3}{21.7}\right)} = -2.97$$

$$(Y - 42.4) = -2.97(x - 6.2) \quad \text{or} \quad Y = 6.8 - 2.97x$$

Significance of b

Residual variance

$$s^2 = 213.3[1-(-0.9459)^2] \times \tfrac{4}{3} = 29.94$$

Standard error of b

$$\epsilon_b = \frac{s}{\sqrt{[\Sigma(x-\bar{x})^2]}} = \sqrt{\left(\frac{29.94}{86.8}\right)} = 0.587$$

Observed value of

$$t = \frac{-2.97-0}{0.587} = -5.06$$

Reference to table 7* for 3 degrees of freedom shows that the value of $|t|$ for significance at the 1% level is 5.841 and at the 2% level is 4.541. The observed value of t falls between the two and it may reasonably be inferred that the slope of the 'true' regression line is different from zero and is negative, the best estimate of its value being -2.97.

Confidence Limits for the Regression Line

The standard error of the regression estimate is

$$\epsilon_{Y_i} = \sqrt{[\epsilon_a^2 + \epsilon_b^2(x_i-\bar{x})^2]} = \sqrt{\left\{29.94\left[\frac{1}{5} + \frac{(x_i-\bar{x})^2}{86.8}\right]\right\}}$$

95% confidence limits for the regression estimate at several values of x_i are derived in table 9.8, figure 9.6 showing these limits plotted on the scatter diagram.

x_i	Y_i	ϵ_{Y_i}	$Y_i - 3.18\,\epsilon_{Y_i}$	$Y_i + 3.18\,\epsilon_{Y_i}$
1	57.8	3.91	45.4	70.2
3	51.9	3.09	42.1	61.7
6.2	42.4	2.45	34.6	50.2
10	31.1	3.31	20.6	41.6
12	25.2	4.19	11.9	38.5

Table 9.8

(b) To estimate the unit cost of an order for eight lenses, substitution of $x = 8$ can be made in the regression equation giving

$$Y = 60.8 - 2.97 \times 8 = £37.0$$

This figure is the 'best' estimate of the average over all possible orders of eight lenses, of the cost per lens in an order of eight lenses.

The uncertainty of this figure (£37.0) is given by the interval (at 95% confidence) £28.50 to £45.50.

If required, the cost per lens for a randomly selected order for eight lenses is likely to be (95% probability) in the interval, £17.64 to £56.36, a very wide range indeed.

5. The scatter diagram (figure 9.7) does not show any evidence against the assumption of linearity and in this example, *a priori* logic suggests that it would be a reasonable model of the situation.

Let x = the number of boxes in a parcel and y = the number of basic seconds per parcel.

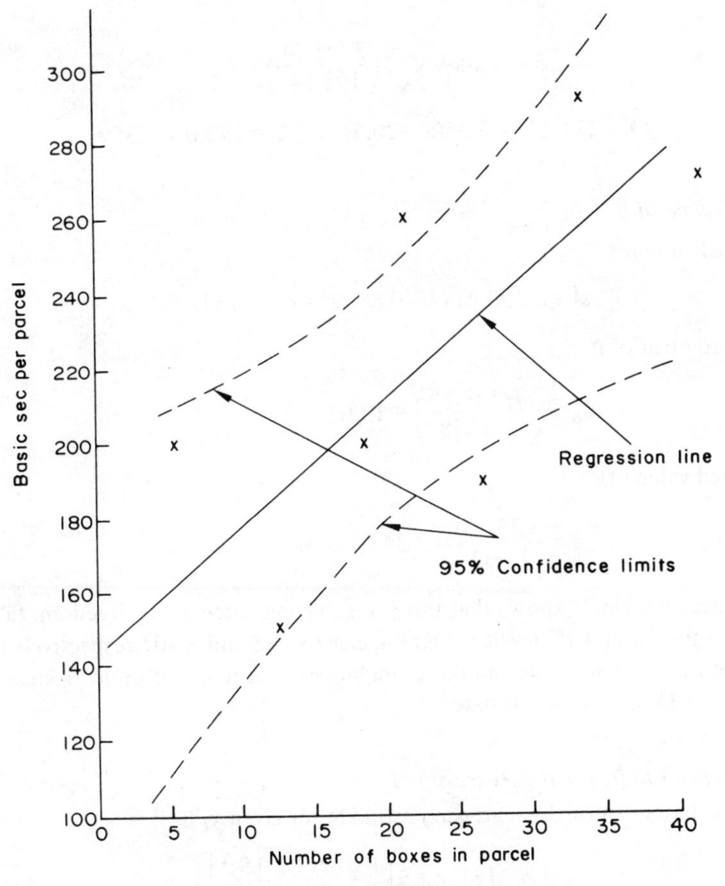

Figure 9.7

The following totals are obtained from the data (without coding)

$$n = 8$$

$\Sigma x = 164$ $\qquad \bar{x} = 20.5$

$\Sigma x^2 = 4700$ $\qquad s_x^2 = 191.14$

$\Sigma y = 1690$ $\qquad \bar{y} = 211.25$

$\Sigma y^2 = 380\,100$ $\qquad s_y^2 = 3298.21$

$\Sigma xy = 39\,130$ $\qquad r = +0.8069$

Regression Line

$$a = \bar{y} = 211.25$$

$$b = 0.8069 \times \sqrt{\left(\frac{3298.21}{191.14}\right)} = 3.35$$

$$(Y - 211.25) = 3.35(x - 20.5) \qquad Y = 142.6 + 3.35x$$

Significance of b

Residual variance

$$s^2 = 3298.21\,(1 - 0.8069^2) \times \tfrac{7}{6} = 1342.5$$

Standard error of b

$$\epsilon_b = \sqrt{\left(\frac{1342.5}{1338}\right)} = 1.002$$

Observed value of

$$t = \frac{3.35 - 0}{1.002} = 3.34$$

Reference to table 7* shows that this value, having 6 degrees of freedom, falls between the 2% and 1% levels of significance (3.143 and 3.707 respectively). The slope of the regression line can therefore be assumed to be different from zero with $b = 3.35$ as its best estimate.

Confidence Limits for the Regression Line

The standard error of the regression estimate for given x_i is

$$\epsilon_{Y_i} = \sqrt{\left\{1342.5\left[\frac{1}{8} + \frac{(x_i - 20.5)^2}{1338}\right]\right\}}$$

Table 9.9 shows values of ϵ_{Y_i} for certain x_i together with 95% confidence limits for the regression estimate at that point. The scatter diagram (figure 9.7) also has 95% confidence limits drawn on it.

x_i	Y_i	ϵ_{Y_i}	$Y_i - 2.45\ \epsilon_{Y_i}$	$Y_i + 2.45\ \epsilon_{Y_i}$
1	145.95	23.44	88.5	203.4
5	159.35	20.22	109.8	208.9
10	176.10	16.69	135.2	217.0
20	209.60	12.96	177.8	241.4
30	243.10	16.07	203.7	282.5
40	276.60	23.44	219.2	334.0

Table 9.9

The analysis therefore gives the following estimates

(a) The constant basic seconds per parcel (i.e. the value of Y at $x = 0$) = 142.6 s and the basic seconds per additional box = 3.35 s.

(b) The *average* time to wrap a parcel with 18 boxes is 202.9 or 203 s.

although the 95% prediction interval for the time taken to wrap a single parcel of 18 boxes is from 107 s to 298 s.

6. Here, in order to reduce the computation slightly, all the basic data have been coded into units of $100; i.e. $1300 becomes 13 etc.

The scatter diagram (figure 9.8) illustrates the case of 'fliers' or 'outliers', i.e. readings which do not appear to belong to the bivariate distribution. These suspect readings are marked as A and B in figure 9.8. Whenever such observations occur in practice, a decision has to be made as to whether or not to exclude them. Special tests to assist in this are available but are beyond the level of this book and all that can be said here is that the source of the readings should be carefully examined and if any reason is found for their not being homogeneous with the others, they should then be rejected. In many cases, a commonsense approach will indicate what should be done.

In this example, the two points, A and B, clearly do not conform and a closer examination of the situation would probably isolate a reason so that the points could validly be excluded. However, to demonstrate their strong effect on the analysis, the points A and B have been retained in fitting the regression line.

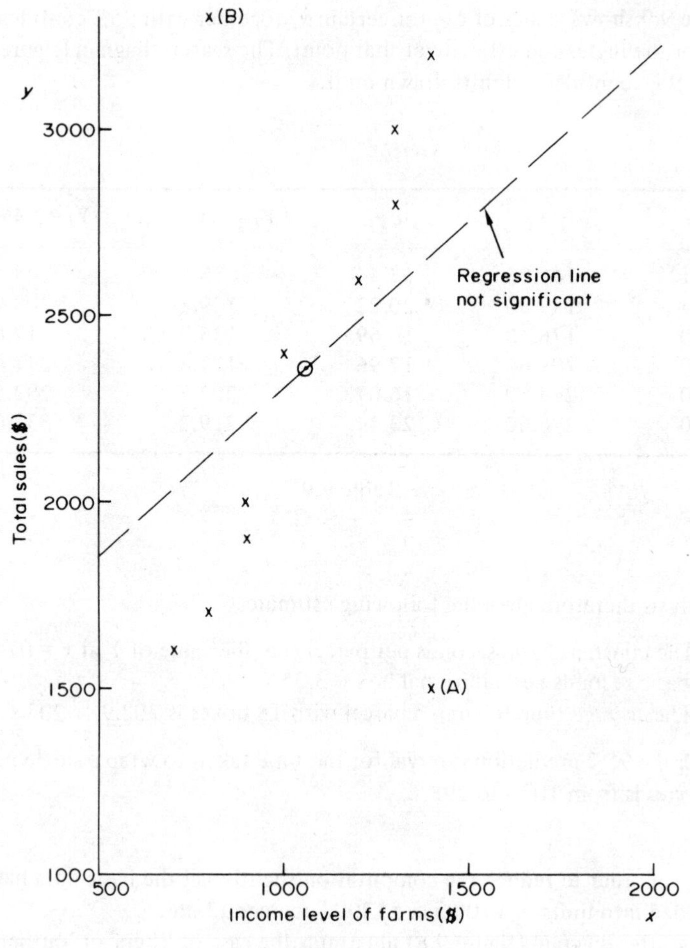

Figure 9.8

x = income level (in \$100)
y = total sales (in \$100)

$n = 11$	$\bar{x} = 10.64$
$\Sigma x = 117$	$s_x^2 = 6.85$
$\Sigma x^2 = 1313$	$\bar{y} = 23.64$
$\Sigma y = 260$	$s_y^2 = 43.45$
$\Sigma y^2 = 6580$	$r = 0.3566$
$\Sigma xy = 2827$	

Regression Line

$$a = \bar{y} = 23.64$$

$$b = 0.3566 \times \sqrt{\left(\frac{43.45}{6.85}\right)} = 0.898$$

$$(Y - 23.64) = 0.898(x - 10.64) \qquad Y = 14.09 + 0.898x \text{ (in \$100)}$$

or converting back to the original units

$$Y = 1409 + 0.898x \text{ (in \$)}$$

Significance of b

The residual variance about the line,

$$s^2 = 43.45(1 - 0.3566^2) \times \tfrac{10}{9} = 42.14$$

Observed

$$t = \frac{b - 0}{\epsilon_b} = \frac{0.898}{0.784} = 1.15$$

a value which is not significantly high.

The regression line calculated above could therefore be misleading since the observed data as a whole show no evidence of a linear relation between y and x.

However, as mentioned above, the analysis can be carried out omitting readings A and B if a valid reason to do so is found. If this is done, the calculations give

$$n = 9$$

$$\Sigma x = 95$$

$$\Sigma x^2 = 1053$$

$$\Sigma y = 212$$

$$\Sigma y^2 = 5266$$

$$\Sigma xy = 2353$$

leading to

$$r = 0.9854 \quad \text{and} \quad Y = 66.15 + 2.29x \text{ (in \$)}$$

The fact that just two points have obscured the relationship should be noted, as should the assistance given by the scatter diagram towards interpretation of the situation.

7. This example illustrates the simple application of regression analysis to time series data.

Note: No attempt will be made to justify forecasting from such analysis (beware of extrapolation) but the method of fitting the linear regression line is given.

As usual, the scatter diagram is plotted and is shown in figure 9.9.

To reduce the size of numbers involved in the computation, the years are coded, 1960 being taken as Year 1 and so on up to 1967 as Year 8.

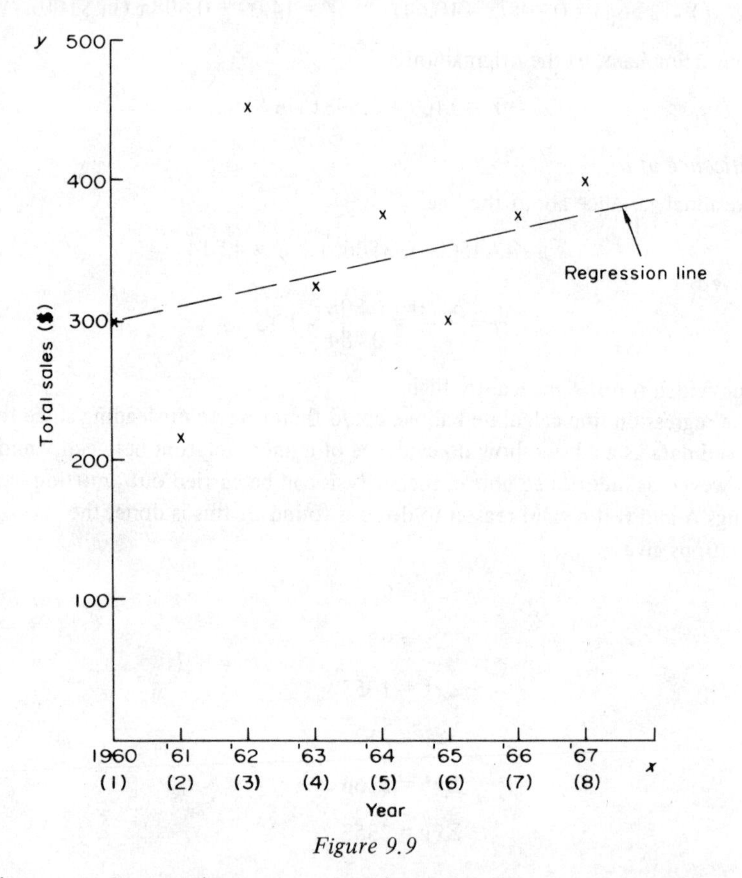

Figure 9.9

The scatter diagram shows no strong relationship between the variables (sales and time), nor is there any apparent evidence of non-linearity, so the results for the straight line regression are as shown below.

$n = 8$	$\Sigma y^2 = 975\,600$	$y = 342.5$
$\Sigma x = 36$	$\Sigma xy = 12\,880$	$s_y^2 = 5307.1$
$\Sigma x^2 = 204$	$\bar{x} = 4.5$	$r = 0.4403$
$\Sigma y = 2740$	$s_x^2 = 6.0$	

Regression Line

$$a = \bar{y} = 342.5$$

$$b = 0.4403\sqrt{\left(\frac{5307.1}{6.0}\right)} = 13.095$$

$$(Y - 342.5) = 13.095(x - 4.5) \qquad Y = 283.57 + 13.09x$$

where x is in coded units.

Significance of b

The residual variance about the line is

$$s^2 = 5307.1[1-(0.4403)^2] \times \tfrac{7}{6} = 4991.3$$

The standard error of b is

$$\epsilon_b = \sqrt{\left(\frac{4991.3}{42}\right)} = 10.90$$

and

$$t = \frac{b - 0}{\epsilon_b} = \frac{13.09}{10.90} = 1.20$$

with 6 degrees of freedom.

Reference to table 7* shows that this is not significantly different from zero, that is, there is no evidence of a relationship between sales and time. In this case there is no point in using the regression equation above to estimate sales for 1968 (Year 9) or beyond. The average yearly sales figure of 342 is probably as good a figure as any to use for making a short-term forecast on the basis of the information given.